Essentials of Statistics

Essentials of Statistics

Edited by
Everett Davies

Larsen & Keller
www.larsen-keller.com

Essentials of Statistics
Edited by Everett Davies
ISBN: 978-1-63549-267-5 (Hardback)

© 2017 Larsen & Keller

▤ Larsen & Keller

Published by Larsen and Keller Education,
5 Penn Plaza,
19th Floor,
New York, NY 10001, USA

Cataloging-in-Publication Data

Essentials of statistics / edited by Everett Davies.
    p. cm.
Includes bibliographical references and index.
ISBN 978-1-63549-267-5
1. Statistics. 2. Mathematical statistics. I. Davies, Everett.
QA276 .E87 2017
001.422--dc23

The publisher's policy is to use permanent paper from mills that operate a sustainable forestry policy. Furthermore, the publisher ensures that the text paper and cover boards used have met acceptable environmental accreditation standards.

Printed and bound in the United States of America.

For more information regarding Larsen and Keller Education and its products, please visit the publisher's website www.larsen-keller.com

Table of Contents

Permissions

Index

Preface

This book elucidates the concepts and innovative models around prospective developments with respect to statistics. It describes in detail the various fundamental methods and uses of this field. Statistics refers to the study of data by organizing, collecting, interpreting, and analyzing it. The various techniques used in mathematical statistics are differential equations, linear algebra, measure-theoretic probability theory, mathematic analysis and stochastic analysis. Some of the diverse topics covered in this text address the varied branches that fall under this category. This book, with its detailed analysis and data prove immensely beneficial to students involved in the field at various levels. For someone with an interest and eye for detail, this textbook covers the most significant topics in the field of mathematics.

Given below is the chapter wise description of the book:

Chapter 1- The organization of data and the collection of data for study is called statistics. It is used in fields that require accurate representation of data for analysis, strategization and decision-making. The following chapter will not only provide an overview, it will also delve deep into the topics related to it.

Chapter 2- The fundamental concepts of statistics are discussed in this chapter. Statistical dispersion, random variable and errors and residuals are some of the significant and important topics related to statistics. The following content unfolds the crucial aspects of statistics in a critical yet systematic manner.

Chapter 3- This chapter explains to the reader the methodologies of statistics. The techniques analyzed in the text are descriptive statistics, statistical inference, univariate analysis, etc. They make possible the various types analysis that this field accomplishes. These aspects are of vital importance, as it provides a better understanding of statistics.

Chapter 4- A central tendency is a concise form of data that represents whole categories of data. Central tendency can be measured in averages. The measures of the subject matter are mean, median and mode and they indicate a different property of the central tendency. The topics discussed in the chapter are of great importance to broaden the existing knowledge on statistics.

Chapter 5- In statistics, deviation is a measure of difference between the value observed and some other value. Some of the topics discussed in this chapter include variance, standard deviation, interquartile range, statistical range and mean absolute difference. The topics discussed in the chapter are of great importance to broaden the existing knowledge on statistical deviation.

Chapter 6- The acquiring of information from a given population is census whereas actuarial science assesses risk in the sectors of finance and insurance. The applications of statistics discussed in this chapter are census, actuarial science, demography, environmental studies etc. The aspects elucidated in this chapter are of vital importance, and provide a better understanding of statistics.

Chapter 7- Statistical methods date back to the 5th century BC. The early applications of statistical thinking dealt with the need of the state to base their policy on the economic data. The modern field of statistics emerged in the late 19th and 20th century. This chapter educates the reader with the history of statistics and progress it has made over a period of time.

At the end, I would like to thank all those who dedicated their time and efforts for the successful completion of this book. I also wish to convey my gratitude towards my friends and family who supported me at every step.

Editor

Introduction to Statistics

The organization of data and the collection of data for study is called statistics. It is used in fields that require accurate representation of data for analysis, strategization and decision-making. The following chapter will not only provide an overview, it will also delve deep into the topics related to it.

Statistics is the study of the collection, analysis, interpretation, presentation, and organization of data. In applying statistics to, e.g., a scientific, industrial, or social problem, it is conventional to begin with a statistical population or a statistical model process to be studied. Populations can be diverse topics such as "all people living in a country" or "every atom composing a crystal". Statistics deals with all aspects of data including the planning of data collection in terms of the design of surveys and experiments.

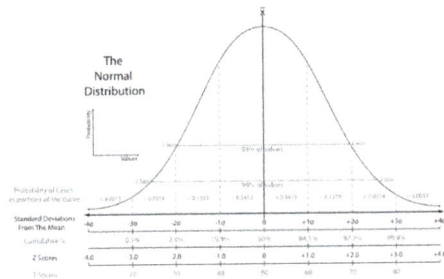

More probability density is found as one gets closer to the expected (mean) value in a normal distribution. Statistics used in standardized testing assessment are shown. The scales include *standard deviations, cumulative percentages, percentile equivalents, Z-scores, T-scores, standard nines,* and *percentages in standard nines.*

Some popular definitions are:

- Merriam-Webster dictionary defines statistics as "classified facts representing the conditions of a people in a state – especially the facts that can be stated in numbers or any other tabular or classified arrangement".

- Statistician Sir Arthur Lyon Bowley defines statistics as "Numerical statements of facts in any department of inquiry placed in relation to each other".

When census data cannot be collected, statisticians collect data by developing specific experiment designs and survey samples. Representative sampling assures that inferences and conclusions can safely extend from the sample to the population as a whole.

An experimental study involves taking measurements of the system under study, manipulating the system, and then taking additional measurements using the same procedure to determine if the manipulation has modified the values of the measurements. In contrast, an observational study does not involve experimental manipulation.

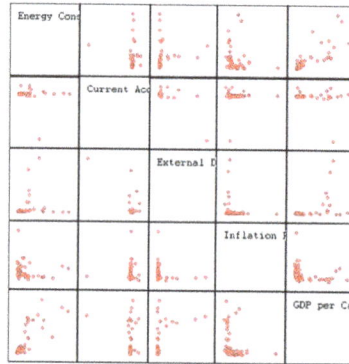

Scatter plots are used in descriptive statistics to show the observed relationships between different variables

Two main statistical methodologies are used in data analysis: descriptive statistics, which summarizes data from a sample using indexes such as the mean or standard deviation, and inferential statistics, which draws conclusions from data that are subject to random variation (e.g., observational errors, sampling variation). Descriptive statistics are most often concerned with two sets of properties of a *distribution* (sample or population): *central tendency* (or *location*) seeks to characterize the distribution's central or typical value, while *dispersion* (or *variability*) characterizes the extent to which members of the distribution depart from its center and each other. Inferences on mathematical statistics are made under the framework of probability theory, which deals with the analysis of random phenomena.

A standard statistical procedure involves the test of the relationship between two statistical data sets, or a data set and a synthetic data drawn from idealized model. An hypothesis is proposed for the statistical relationship between the two data sets, and this is compared as an alternative to an idealized null hypothesis of no relationship between two data sets. Rejecting or disproving the null hypothesis is done using statistical tests that quantify the sense in which the null can be proven false, given the data that are used in the test. Working from a null hypothesis, two basic forms of error are recognized: Type I errors (null hypothesis is falsely rejected giving a "false positive") and Type II errors (null hypothesis fails to be rejected and an actual difference between populations is missed giving a "false negative"). Multiple problems have come to be associated with this framework: ranging from obtaining a sufficient sample size to specifying an adequate null hypothesis.

Measurement processes that generate statistical data are also subject to error. Many of these errors are classified as random (noise) or systematic (bias), but other types of errors (e.g., blunder, such as when an analyst reports incorrect units) can also be important. The presence of missing data and/or censoring may result in biased estimates and specific techniques have been developed to address these problems.

Statistics can be said to have begun in ancient civilization, going back at least to the 5th century BC, but it was not until the 18th century that it started to draw more heavily from calculus and probability theory. Statistics continues to be an area of active research, for example on the problem of how to analyze Big data.

Scope

Statistics is a mathematical body of science that pertains to the collection, analysis, interpretation or explanation, and presentation of data,or as a branch of mathematics. Some consider statistics to be a distinct mathematical science rather than a branch of mathematics. While many scientific investigations make use of data, statistics is concerned with the use of data in the context of uncertainty and decision making in the face of uncertainty.

Mathematical Statistics

Mathematical statistics is the application of mathematics to statistics, which was originally conceived as the science of the state — the collection and analysis of facts about a country: its economy, land, military, population, and so forth. Mathematical techniques used for this include mathematical analysis, linear algebra, stochastic analysis, differential equations, and measure-theoretic probability theory.

Overview

In applying statistics to a problem, it is common practice to start with a population or process to be studied. Populations can be diverse topics such as "all persons living in a country" or "every atom composing a crystal".

Ideally, statisticians compile data about the entire population (an operation called census). This may be organized by governmental statistical institutes. Descriptive statistics can be used to summarize the population data. Numerical descriptors include mean and standard deviation for continuous data types (like income), while frequency and percentage are more useful in terms of describing categorical data (like race).

When a census is not feasible, a chosen subset of the population called a sample is studied. Once a sample that is representative of the population is determined, data is collected for the sample members in an observational or experimental setting. Again, descriptive statistics can be used to summarize the sample data. However, the drawing of the sample has been subject to an element of randomness, hence the established numerical descriptors from the sample are also due to uncertainty. To still draw meaningful conclusions about the entire population, inferential statistics is needed. It uses patterns in the sample data to draw inferences about the population represented, accounting for randomness. These inferences may take the form of: answering yes/no questions about the data (hypothesis testing), estimating numerical characteristics of the data (estimation), describing associations within the data (correlation) and modeling relationships within the data (for example, using regression analysis). Inference can extend to forecasting, prediction and estimation of unobserved values either in or associated with the population being studied; it can include extrapolation and interpolation of time series or spatial data, and can also include data mining.

Data Collection

Sampling

When full census data cannot be collected, statisticians collect sample data by developing specific experiment designs and survey samples. Statistics itself also provides tools for prediction and forecasting the use of data through statistical models. To use a sample as a guide to an entire population, it is important that it truly represents the overall population. Representative sampling assures that inferences and conclusions can safely extend from the sample to the population as a whole. A major problem lies in determining the extent that the sample chosen is actually representative. Statistics offers methods to estimate and correct for any bias within the sample and data collection procedures. There are also methods of experimental design for experiments that can lessen these issues at the outset of a study, strengthening its capability to discern truths about the population.

Sampling theory is part of the mathematical discipline of probability theory. Probability is used in mathematical statistics to study the sampling distributions of sample statistics and, more generally, the properties of statistical procedures. The use of any statistical method is valid when the system or population under consideration satisfies the assumptions of the method. The difference in point of view between classic probability theory and sampling theory is, roughly, that probability theory starts from the given parameters of a total population to deduce probabilities that pertain to samples. Statistical inference, however, moves in the opposite direction—inductively inferring from samples to the parameters of a larger or total population.

Experimental and Observational Studies

A common goal for a statistical research project is to investigate causality, and in particular to draw a conclusion on the effect of changes in the values of predictors or independent variables on dependent variables. There are two major types of causal statistical studies: experimental studies and observational studies. In both types of studies, the effect of differences of an independent variable (or variables) on the behavior of the dependent variable are observed. The difference between the two types lies in how the study is actually conducted. Each can be very effective. An experimental study involves taking measurements of the system under study, manipulating the system, and then taking additional measurements using the same procedure to determine if the manipulation has modified the values of the measurements. In contrast, an observational study does not involve experimental manipulation. Instead, data are gathered and correlations between predictors and response are investigated. While the tools of data analysis work best on data from randomized studies, they are also applied to other kinds of data – like natural experiments and observational studies – for which a statistician would use a modified, more structured estimation method (e.g., Difference in differences estimation and instrumental variables, among many others) that produce consistent estimators.

Experiments

The basic steps of a statistical experiment are:

1. Planning the research, including finding the number of replicates of the study, using the following information: preliminary estimates regarding the size of treatment effects, alternative hypotheses, and the estimated experimental variability. Consideration of the selection of experimental subjects and the ethics of research is necessary. Statisticians recommend that experiments compare (at least) one new treatment with a standard treatment or control, to allow an unbiased estimate of the difference in treatment effects.

2. Design of experiments, using blocking to reduce the influence of confounding variables, and randomized assignment of treatments to subjects to allow unbiased estimates of treatment effects and experimental error. At this stage, the experimenters and statisticians write the *experimental protocol* that will guide the performance of the experiment and which specifies the *primary analysis* of the experimental data.

3. Performing the experiment following the experimental protocol and analyzing the data following the experimental protocol.

4. Further examining the data set in secondary analyses, to suggest new hypotheses for future study.

5. Documenting and presenting the results of the study.

Experiments on human behavior have special concerns. The famous Hawthorne study examined changes to the working environment at the Hawthorne plant of the Western Electric Company. The researchers were interested in determining whether increased illumination would increase the productivity of the assembly line workers. The researchers first measured the productivity in the plant, then modified the illumination in an area of the plant and checked if the changes in illumination affected productivity. It turned out that productivity indeed improved (under the experimental conditions).

However, the study is heavily criticized today for errors in experimental procedures, specifically for the lack of a control group and blindness. The Hawthorne effect refers to finding that an outcome (in this case, worker productivity) changed due to observation itself. Those in the Hawthorne study became more productive not because the lighting was changed but because they were being observed.

Observational Study

An example of an observational study is one that explores the association between smoking and lung cancer. This type of study typically uses a survey to collect observations about the area of interest and then performs statistical analysis. In this case, the researchers would collect observations of both smokers and non-smokers, perhaps through a case-control study, and then look for the number of cases of lung cancer in each group.

Types of Data

Various attempts have been made to produce a taxonomy of levels of measurement. The psychophysicist Stanley Smith Stevens defined nominal, ordinal, interval, and ratio scales. Nominal measurements do not have meaningful rank order among values, and permit any one-to-one transformation. Ordinal measurements have imprecise differences between consecutive values, but have a meaningful order to those values, and permit any order-preserving transformation. Interval measurements have meaningful distances between measurements defined, but the zero value is arbitrary (as in the case with longitude and temperature measurements in Celsius or Fahrenheit), and permit any linear transformation. Ratio measurements have both a meaningful zero value and the distances between different measurements defined, and permit any rescaling transformation.

Because variables conforming only to nominal or ordinal measurements cannot be reasonably measured numerically, sometimes they are grouped together as categorical variables, whereas ratio and interval measurements are grouped together as quantitative variables, which can be either discrete or continuous, due to their numerical nature. Such distinctions can often be loosely correlated with data type in computer science, in that dichotomous categorical variables may be represented with the Boolean data type, polytomous categorical variables with arbitrarily assigned integers in the integral data type, and continuous variables with the real data type involving floating point computation. But the mapping of computer science data types to statistical data types depends on which categorization of the latter is being implemented.

Other categorizations have been proposed. For example, Mosteller and Tukey (1977) distinguished grades, ranks, counted fractions, counts, amounts, and balances. Nelder (1990) described continuous counts, continuous ratios, count ratios, and categorical modes of data.

The issue of whether or not it is appropriate to apply different kinds of statistical methods to data obtained from different kinds of measurement procedures is complicated by issues concerning the transformation of variables and the precise interpretation of research questions. "The relationship between the data and what they describe merely reflects the fact that certain kinds of statistical statements may have truth values which are not invariant under some transformations. Whether or not a transformation is sensible to contemplate depends on the question one is trying to answer" (Hand, 2004, p. 82).

Terminology and Theory of Inferential Statistics

Statistics, Estimators and Pivotal Quantities

Consider independent identically distributed (IID) random variables with a given probability distribution: standard statistical inference and estimation theory defines a random sample as the random vector given by the column vector of these IID variables. The population being examined is described by a probability distribution that may have unknown parameters.

A statistic is a random variable that is a function of the random sample, but *not a function of unknown parameters*. The probability distribution of the statistic, though, may have unknown parameters.

Consider now a function of the unknown parameter: an estimator is a statistic used to estimate such function. Commonly used estimators include sample mean, unbiased sample variance and sample covariance.

A random variable that is a function of the random sample and of the unknown parameter, but whose probability distribution *does not depend on the unknown parameter* is called a pivotal quantity or pivot. Widely used pivots include the z-score, the chi square statistic and Student's t-value.

Between two estimators of a given parameter, the one with lower mean squared error is said to be more efficient. Furthermore, an estimator is said to be unbiased if its expected value is equal to the true value of the unknown parameter being estimated, and asymptotically unbiased if its expected value converges at the limit to the true value of such parameter.

Other desirable properties for estimators include: UMVUE estimators that have the lowest variance for all possible values of the parameter to be estimated (this is usually an easier property to verify than efficiency) and consistent estimators which converges in probability to the true value of such parameter.

This still leaves the question of how to obtain estimators in a given situation and carry the computation, several methods have been proposed: the method of moments, the maximum likelihood method, the least squares method and the more recent method of estimating equations.

Null hypothesis and Alternative Hypothesis

Interpretation of statistical information can often involve the development of a null hypothesis which is usually (but not necessarily) that no relationship exists among variables or that no change occurred over time.

The best illustration for a novice is the predicament encountered by a criminal trial. The null hypothesis, H_0, asserts that the defendant is innocent, whereas the alternative hypothesis, H_1, asserts that the defendant is guilty. The indictment comes because of suspicion of the guilt. The H_0 (status quo) stands in opposition to H_1 and is maintained unless H_1 is supported by evidence "beyond a reasonable doubt". However, "failure to reject H_0" in this case does not imply innocence, but merely that the evidence was insufficient to convict. So the jury does not necessarily *accept* H_0 but *fails to reject* H_0. While one can not "prove" a null hypothesis, one can test how close it is to being true with a power test, which tests for type II errors.

What statisticians call an alternative hypothesis is simply an hypothesis that contradicts the null hypothesis.

Error

Working from a null hypothesis, two basic forms of error are recognized:

- Type I errors where the null hypothesis is falsely rejected giving a "false positive".

- Type II errors where the null hypothesis fails to be rejected and an actual difference between populations is missed giving a "false negative".

Standard deviation refers to the extent to which individual observations in a sample differ from a central value, such as the sample or population mean, while Standard error refers to an estimate of difference between sample mean and population mean.

A statistical error is the amount by which an observation differs from its expected value, a residual is the amount an observation differs from the value the estimator of the expected value assumes on a given sample (also called prediction).

Mean squared error is used for obtaining efficient estimators, a widely used class of estimators. Root mean square error is simply the square root of mean squared error.

Many statistical methods seek to minimize the residual sum of squares, and these are called "methods of least squares" in contrast to Least absolute deviations. The latter gives equal weight to small and big errors, while the former gives more weight to large errors. Residual sum of squares is also differentiable, which provides a handy property for doing regression. Least squares applied to linear regression is called ordinary least squares method and least squares applied to nonlinear regression is called non-linear least squares. Also in a linear regression model the non deterministic part of the model is called error term, disturbance or more simply noise. Both linear regression and non-linear regression are addressed in polynomial least squares, which also describes the variance in a prediction of the dependent variable (y axis) as a function of the independent variable (x axis) and the deviations (errors, noise, disturbances) from the estimated (fitted) curve.

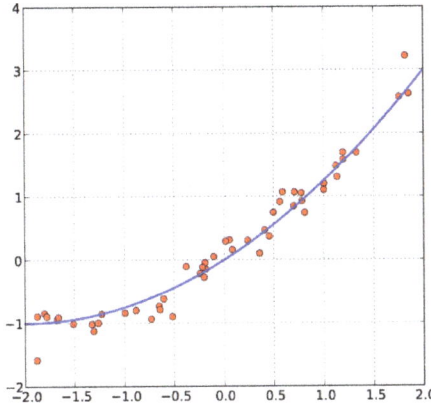

A least squares fit: in red the points to be fitted, in blue the fitted line.

Measurement processes that generate statistical data are also subject to error. Many of these errors are classified as random (noise) or systematic (bias), but other types of errors (e.g., blunder, such as when an analyst reports incorrect units) can also be important. The presence of missing data and/or censoring may result in biased estimates and specific techniques have been developed to address these problems.

Interval Estimation

Confidence intervals: the red line is true value for the mean in this example, the blue lines are random confidence intervals for 100 realizations.

Most studies only sample part of a population, so results don't fully represent the whole population. Any estimates obtained from the sample only approximate the population value. Confidence intervals allow statisticians to express how closely the sample estimate matches the true value in the whole population. Often they are expressed as 95% confidence intervals. Formally, a 95% confidence interval for a value is a range where, if the sampling and analysis were repeated under the same conditions (yielding a different dataset), the interval would include the true (population) value in 95% of all possible cases. This does *not* imply that the probability that the true value is in the confidence interval is 95%. From the frequentist perspective, such a claim does not even make sense, as the true value is not a random variable. Either the true value is or is not within the given interval. However, it is true that, before any data are sampled and given a plan for how to construct the confidence interval, the probability is 95% that the yet-to-be-calculated interval will cover the true value: at this point, the limits of the interval are yet-to-be-observed random variables. One approach that does yield an interval that can be interpreted as having a given probability of containing the true value is to use a credible interval from Bayesian statistics: this approach depends on a different way of interpreting what is meant by "probability", that is as a Bayesian probability.

In principle confidence intervals can be symmetrical or asymmetrical. An interval can be asymmetrical because it works as lower or upper bound for a parameter (left-sided interval or right sided interval), but it can also be asymmetrical because the two sided interval is built violating symmetry around the estimate. Sometimes the bounds for a confidence interval are reached asymptotically and these are used to approximate the true bounds.

Significance

Statistics rarely give a simple Yes/No type answer to the question under analysis. Interpretation often comes down to the level of statistical significance applied to the numbers and often refers to the probability of a value accurately rejecting the null hypothesis (sometimes referred to as the p-value).

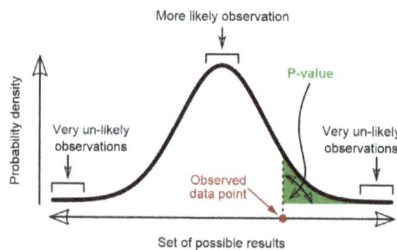

Important:

Pr (observation | hypothesis) ≠ Pr (hypothesis | observation)

The probability of observing a result given that some hypothesis is true is *not equivalent* to the probability that a hypothesis is true given that some result has been observed.

Using the p-value as a "score" is committing an egregious logical error: **the transposed conditional fallacy**.

A p-value (shaded green area) is the probability of an observed (or more extreme) result assuming that the null hypothesis is true.

In this graph the black line is probability distribution for the test statistic, the critical region is the set of values to the right of the observed data point (observed value of the test statistic) and the p-value is represented by the green area.

The standard approach is to test a null hypothesis against an alternative hypothesis. A critical region is the set of values of the estimator that leads to refuting the null hypothesis. The probability of type I error is therefore the probability that the estimator belongs to the critical region given that null hypothesis is true (statistical significance) and the probability of type II error is the probability that the estimator doesn't belong to the critical region given that the alternative hypothesis is true. The statistical power of a test is the probability that it correctly rejects the null hypothesis when the null hypothesis is false.

Referring to statistical significance does not necessarily mean that the overall result is significant in real world terms. For example, in a large study of a drug it may be shown that the drug has a statistically significant but very small beneficial effect, such that the drug is unlikely to help the patient noticeably.

While in principle the acceptable level of statistical significance may be subject to debate, the p-value is the smallest significance level that allows the test to reject the null hypothesis. This is logically equivalent to saying that the p-value is the probability, assuming the null hypothesis is true, of observing a result at least as extreme as the test statistic. Therefore, the smaller the p-value, the lower the probability of committing type I error.

Some problems are usually associated with this framework:

- A difference that is highly statistically significant can still be of no practical significance, but it is possible to properly formulate tests to account for this. One response involves going beyond reporting only the significance level to include the *p*-value when reporting whether a hypothesis is rejected or accepted. The p-value, however, does not indicate the size or importance of the observed effect and can also seem to exaggerate the importance of minor differences in large studies. A better and increasingly common approach is to report confidence intervals. Although these are produced from the same calculations as those of hypothesis tests or *p*-values, they describe both the size of the effect and the uncertainty surrounding it.

- Fallacy of the transposed conditional, aka prosecutor's fallacy: criticisms arise because the hypothesis testing approach forces one hypothesis (the null hypothesis) to be favored, since what is being evaluated is probability of the observed result given the null hypothesis and not probability of the null hypothesis given the observed result. An alternative to this approach is offered by Bayesian inference, although it requires establishing a prior probability.

- Rejecting the null hypothesis does not automatically prove the alternative hypothesis.

- As everything in inferential statistics it relies on sample size, and therefore under fat tails p-values may be seriously mis-computed.

Examples

Some well-known statistical tests and procedures are:

- Analysis of variance (ANOVA)

- Chi-squared test

- Correlation

- Factor analysis

- Mann–Whitney U

- Mean square weighted deviation (MSWD)

- Pearson product-moment correlation coefficient

- Regression analysis

- Spearman's rank correlation coefficient

- Student's t-test

- Time series analysis

- Conjoint Analysis

Misuse of Statistics

Misuse of statistics can produce subtle, but serious errors in description and interpretation—subtle in the sense that even experienced professionals make such errors, and serious in the sense that they can lead to devastating decision errors. For instance, social policy, medical practice, and the reliability of structures like bridges all rely on the proper use of statistics.

Even when statistical techniques are correctly applied, the results can be difficult to interpret for those lacking expertise. The statistical significance of a trend in the data—which measures the extent to which a trend could be caused by random variation in the sample—may or may not agree with an intuitive sense of its significance. The set of basic statistical skills (and skepticism) that people need to deal with information in their everyday lives properly is referred to as statistical literacy.

There is a general perception that statistical knowledge is all-too-frequently intentionally misused by finding ways to interpret only the data that are favorable to the presenter. A mistrust and misunderstanding of statistics is associated with the quotation, "There are three kinds of lies: lies, damned lies, and statistics". Misuse of statistics can be both inadvertent and intentional, and the book *How to Lie with Statistics* outlines a range of considerations. In an attempt to shed light on the use and misuse of statistics, reviews of statistical techniques used in particular fields are conducted (e.g. Warne, Lazo, Ramos, and Ritter (2012)).

Ways to avoid misuse of statistics include using proper diagrams and avoiding bias. Misuse can occur when conclusions are overgeneralized and claimed to be representative of more than they really are, often by either deliberately or unconsciously overlooking sampling bias. Bar graphs are arguably the easiest diagrams to use and understand, and they can be made either by hand or with simple computer programs. Unfortunately, most people do not look for bias or errors, so they are not noticed. Thus, people may often believe that something is true even if it is not well represented. To make data gathered from statistics believable and accurate, the sample taken must be representative of the whole. According to Huff, "The dependability of a sample can be destroyed by [bias]... allow yourself some degree of skepticism."

To assist in the understanding of statistics Huff proposed a series of questions to be asked in each case:

- Who says so? (Does he/she have an axe to grind?)

- How does he/she know? (Does he/she have the resources to know the facts?)

- What's missing? (Does he/she give us a complete picture?)

- Did someone change the subject? (Does he/she offer us the right answer to the wrong problem?)

- Does it make sense? (Is his/her conclusion logical and consistent with what we already know?)

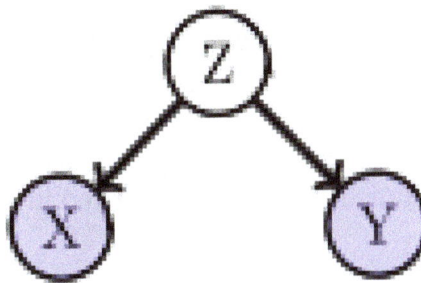

The confounding variable problem: X and Y may be correlated, not because there is causal relationship between them, but because both depend on a third variable Z. Z is called a confounding factor.

Misinterpretation: Correlation

The concept of correlation is particularly noteworthy for the potential confusion it can cause. Statistical analysis of a data set often reveals that two variables (properties) of the population under consideration tend to vary together, as if they were connected. For example, a study of annual income that also looks at age of death might find that poor people tend to have shorter lives than affluent people. The two variables are said to be correlated; however, they may or may not be the cause of one another. The correlation phenomena could be caused by a third, previously unconsidered phenomenon, called a lurking variable or confounding variable. For this reason, there is no way to immediately infer the existence of a causal relationship between the two variables.

History of Statistical Science

Gerolamo Cardano, the earliest pioneer on the mathematics of probability.

Statistical methods date back at least to the 5th century BC.

Some scholars pinpoint the origin of statistics to 1663, with the publication of *Natural and Political Observations upon the Bills of Mortality* by John Graunt. Early applications of statistical thinking revolved around the needs of states to base policy on demographic and economic data, hence its *stat-* etymology. The scope of the discipline of statistics broadened in the early 19th century to include the collection and analysis of data in general. Today, statistics is widely employed in government, business, and natural and social sciences.

Its mathematical foundations were laid in the 17th century with the development of the probability theory by Gerolamo Cardano, Blaise Pascal and Pierre de Fermat. Mathematical probability theory arose from the study of games of chance, although the concept of probability was already examined in medieval law and by philosophers such as Juan Caramuel. The method of least squares was first described by Adrien-Marie Legendre in 1805.

The modern field of statistics emerged in the late 19th and early 20th century in three stages. The first wave, at the turn of the century, was led by the work of Francis Galton and Karl Pearson, who transformed statistics into a rigorous mathematical discipline used for analysis, not just in science, but in industry and politics as well. Galton's contributions included introducing the concepts of standard deviation, correlation, regression analysis and the application of these methods to the study of the variety of human characteristics – height, weight, eyelash length among others. Pearson developed the Pearson product-moment correlation coefficient, defined as a product-moment, the method of moments for the fitting of distributions to samples and the Pearson distribution, among many other things. Galton and Pearson founded *Biometrika* as the first journal of mathematical statistics and biostatistics (then called biometry), and the latter founded the world's first university statistics department at University College London.

Karl Pearson, a founder of mathematical statistics.

The modern field of statistics emerged in the late 19th and early 20th century in three stages. The first wave, at the turn of the century, was led by the work of Francis Galton and Karl Pearson, who transformed statistics into a rigorous mathematical discipline used for analysis, not just in science, but in industry and politics as well. Galton's contributions included introducing the concepts of standard deviation, correlation, regression analysis and the application of these methods to the study of the variety of human characteristics – height, weight, eyelash length among others. Pearson developed the Pearson product-moment correlation coefficient, defined as a product-moment, the method of moments for the fitting of distributions to samples and the Pearson distribution, among many other things. Galton and Pearson founded *Biometrika* as the first journal of mathematical statistics and biostatistics (then called biometry), and the latter founded the world's first university statistics department at University College London.

Ronald Fisher coined the term null hypothesis during the Lady tasting tea experiment, which "is never proved or established, but is possibly disproved, in the course of experimentation".

The second wave of the 1910s and 20s was initiated by William Gosset, and reached its culmination in the insights of Ronald Fisher, who wrote the textbooks that were to define the academic discipline in universities around the world. Fisher's most important publications were his 1918 seminal paper *The Correlation between Relatives on the*

Supposition of Mendelian Inheritance, which was the first to use the statistical term, variance, his classic 1925 work *Statistical Methods for Research Workers* and his 1935 *The Design of Experiments*, where he developed rigorous design of experiments models. He originated the concepts of sufficiency, ancillary statistics, Fisher's linear discriminator and Fisher information. In his 1930 book *The Genetical Theory of Natural Selection* he applied statistics to various biological concepts such as Fisher's principle). Nevertheless, A. W. F. Edwards has remarked that it is "probably the most celebrated argument in evolutionary biology". (about the sex ratio), the Fisherian runaway, a concept in sexual selection about a positive feedback runaway affect found in evolution.

The final wave, which mainly saw the refinement and expansion of earlier developments, emerged from the collaborative work between Egon Pearson and Jerzy Neyman in the 1930s. They introduced the concepts of "Type II" error, power of a test and confidence intervals. Jerzy Neyman in 1934 showed that stratified random sampling was in general a better method of estimation than purposive (quota) sampling.

Today, statistical methods are applied in all fields that involve decision making, for making accurate inferences from a collated body of data and for making decisions in the face of uncertainty based on statistical methodology. The use of modern computers has expedited large-scale statistical computations, and has also made possible new methods that are impractical to perform manually. Statistics continues to be an area of active research, for example on the problem of how to analyze Big data.

Applications

Applied Statistics, Theoretical Statistics and Mathematical Statistics

"Applied statistics" comprises descriptive statistics and the application of inferential statistics. *Theoretical statistics* concerns both the logical arguments underlying justification of approaches to statistical inference, as well encompassing *mathematical statistics*. Mathematical statistics includes not only the manipulation of probability distributions necessary for deriving results related to methods of estimation and inference, but also various aspects of computational statistics and the design of experiments.

Machine Learning and Data Mining

There are two applications for machine learning and data mining: data management and data analysis. Statistics tools are necessary for the data analysis.

Statistics in Society

Statistics is applicable to a wide variety of academic disciplines, including natural and social sciences, government, and business. Statistical consultants can help organizations and companies that don't have in-house expertise relevant to their particular questions.

Statistical Computing

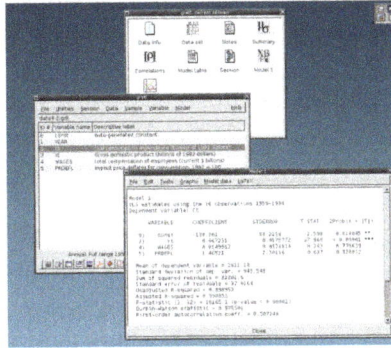

gretl, an example of an open source statistical package

The rapid and sustained increases in computing power starting from the second half of the 20th century have had a substantial impact on the practice of statistical science. Early statistical models were almost always from the class of linear models, but powerful computers, coupled with suitable numerical algorithms, caused an increased interest in nonlinear models (such as neural networks) as well as the creation of new types, such as generalized linear models and multilevel models.

Increased computing power has also led to the growing popularity of computationally intensive methods based on resampling, such as permutation tests and the bootstrap, while techniques such as Gibbs sampling have made use of Bayesian models more feasible. The computer revolution has implications for the future of statistics with new emphasis on "experimental" and "empirical" statistics. A large number of both general and special purpose statistical software are now available.

Statistics Applied to Mathematics or the Arts

Traditionally, statistics was concerned with drawing inferences using a semi-standardized methodology that was "required learning" in most sciences. This has changed with use of statistics in non-inferential contexts. What was once considered a dry subject, taken in many fields as a degree-requirement, is now viewed enthusiastically. Initially derided by some mathematical purists, it is now considered essential methodology in certain areas.

- In number theory, scatter plots of data generated by a distribution function may be transformed with familiar tools used in statistics to reveal underlying patterns, which may then lead to hypotheses.

- Methods of statistics including predictive methods in forecasting are combined with chaos theory and fractal geometry to create video works that are considered to have great beauty.

- The process art of Jackson Pollock relied on artistic experiments whereby underlying distributions in nature were artistically revealed. With the advent of computers, statistical methods were applied to formalize such distribution-driven natural processes to make and analyze moving video art.

- Methods of statistics may be used predicatively in performance art, as in a card trick based on a Markov process that only works some of the time, the occasion of which can be predicted using statistical methodology.

- Statistics can be used to predicatively create art, as in the statistical or stochastic music invented by Iannis Xenakis, where the music is performance-specific. Though this type of artistry does not always come out as expected, it does behave in ways that are predictable and tunable using statistics.

Specialized Disciplines

Statistical techniques are used in a wide range of types of scientific and social research, including: biostatistics, computational biology, computational sociology, network biology, social science, sociology and social research. Some fields of inquiry use applied statistics so extensively that they have specialized terminology. These disciplines include:

- Actuarial science (assesses risk in the insurance and finance industries)

- Applied information economics

- Astrostatistics (statistical evaluation of astronomical data)

- Biostatistics

- Business statistics

- Chemometrics (for analysis of data from chemistry)

- Data mining (applying statistics and pattern recognition to discover knowledge from data)

- Data science

- Demography

- Econometrics (statistical analysis of economic data)

- Energy statistics

- Engineering statistics

- Epidemiology (statistical analysis of disease)

- Geography and Geographic Information Systems, specifically in Spatial analysis

- Image processing

- Medical Statistics

- Psychological statistics

- Reliability engineering

- Social statistics

- Statistical Mechanics

In addition, there are particular types of statistical analysis that have also developed their own specialised terminology and methodology:

- Bootstrap / Jackknife resampling

- Multivariate statistics

- Statistical classification

- Structured data analysis (statistics)

- Structural equation modelling

- Survey methodology

- Survival analysis

- Statistics in various sports, particularly baseball - known as Sabermetrics - and cricket

Statistics form a key basis tool in business and manufacturing as well. It is used to understand measurement systems variability, control processes (as in statistical process control or SPC), for summarizing data, and to make data-driven decisions. In these roles, it is a key tool, and perhaps the only reliable tool.

References

- Hays, William Lee, (1973) Statistics for the Social Sciences, Holt, Rinehart and Winston, p.xii, ISBN 978-0-03-077945-9

- Chance, Beth L.; Rossman, Allan J. (2005). "Preface". Investigating Statistical Concepts, Applications, and Methods (PDF). Duxbury Press. ISBN 978-0-495-05064-3.

- Lakshmikantham,, ed. by D. Kannan,... V. (2002). Handbook of stochastic analysis and applications. New York: M. Dekker. ISBN 0824706609.

- Freedman, D.A. (2005) Statistical Models: Theory and Practice, Cambridge University Press. ISBN 978-0-521-67105-7

- Everitt, Brian (1998). The Cambridge Dictionary of Statistics. Cambridge, UK New York: Cambridge University Press. ISBN 0521593468.

- Huff, Darrell (1954) How to Lie with Statistics, WW Norton & Company, Inc. New York, NY. ISBN 0-393-31072-8

- Drennan, Robert D. (2008). "Statistics in archaeology". In Pearsall, Deborah M. Encyclopedia of Archaeology. Elsevier Inc. pp. 2093–2100. ISBN 978-0-12-373962-9.

- Anderson, D.R.; Sweeney, D.J.; Williams, T.A. (1994) Introduction to Statistics: Concepts and Applications, pp. 5–9. West Group. ISBN 978-0-314-03309-3

- "Essay on Statistics: Meaning and Definition of Statistics". Economics Discussion. 2014-12-02. Retrieved 2016-05-28.

- "What Is the Difference Between Type I and Type II Hypothesis Testing Errors?". About.com Education. Retrieved 2015-11-27.

Essential Concepts of Statistics

The fundamental concepts of statistics are discussed in this chapter. Statistical dispersion, random variable and errors and residuals are some of the significant and important topics related to statistics. The following content unfolds the crucial aspects of statistics in a critical yet systematic manner.

Statistical Dispersion

In statistics, dispersion (also called variability, scatter, or spread) denotes how stretched or squeezed a distribution (theoretical or that underlying a statistical sample) is. Common examples of measures of statistical dispersion are the variance, standard deviation and interquartile range.

Dispersion is contrasted with location or central tendency, and together they are the most used properties of distributions.

Measures of Statistical Dispersion

A measure of statistical dispersion is a nonnegative real number that is zero if all the data are the same and increases as the data become more diverse.

Most measures of dispersion have the same units as the quantity being measured. In other words, if the measurements are in metres or seconds, so is the measure of dispersion. Such measures of dispersion include:

- Sample standard deviation
- Interquartile range (IQR)
- Range
- Mean absolute difference (also known as Gini mean absolute difference)
- Median absolute deviation (MAD)
- Average absolute deviation (or simply called average deviation)
- Distance standard deviation

These are frequently used (together with scale factors) as estimators of scale parameters, in which capacity they are called estimates of scale. Robust measures of scale are those unaffected by a small number of outliers, and include the IQR and MAD.

All the above measures of statistical dispersion have the useful property that they are location-invariant and linear in scale. This means that if a random variable X has a dispersion of S_X then a linear transformation $Y = aX + b$ for real a and b should have dispersion $S_Y = |a|S_X$, where $|a|$ is the absolute value of a, that is, ignores a preceding negative sign –.

Other measures of dispersion are dimensionless. In other words, they have no units even if the variable itself has units. These include:

- Coefficient of variation

- Quartile coefficient of dispersion

- Relative mean difference, equal to twice the Gini coefficient

- Entropy: While the entropy of a discrete variable is location-invariant and scale-independent, and therefore not a measure of dispersion in the above sense, the entropy of a continuous variable is location invariant and additive in scale: If Hz is the entropy of continuous variable z and $y=ax+b$, then $Hy=Hx-+log(a)$.

There are other measures of dispersion:

- Variance (the square of the standard deviation) – location-invariant but not linear in scale.

- Variance-to-mean ratio – mostly used for count data when the term coefficient of dispersion is used and when this ratio is dimensionless, as count data are themselves dimensionless, not otherwise.

Some measures of dispersion have specialized purposes, among them the Allan variance and the Hadamard variance.

For categorical variables, it is less common to measure dispersion by a single number. One measure that does so is the discrete entropy.

Sources of Statistical Dispersion

In the physical sciences, such variability may result from random measurement errors: instrument measurements are often not perfectly precise, i.e., reproducible, and there is additional inter-rater variability in interpreting and reporting the measured results. One may assume that the quantity being measured is stable, and that the variation between measurements is due to observational error. A system of a large number of particles is characterized by the mean values of a relatively few number of macroscopic quantities such as temperature, energy, and density. The standard deviation is an important measure in Fluctuation theory, which explains many physical phenomena, including why the sky is blue.

In the biological sciences, the quantity being measured is seldom unchanging and stable, and the variation observed might additionally be *intrinsic* to the phenomenon: It may be due to *inter-individual variability*, that is, distinct members of a population differing from each other. Also, it may be due to *intra-individual variability*, that is, one and the same subject differing in tests taken at different times or in other differing conditions. Such types of variability are also seen in the arena of manufactured products; even there, the meticulous scientist finds variation.

In economics, finance, and other disciplines, regression analysis attempts to explain the dispersion of a dependent variable, generally measured by its variance, using one or more independent variables each of which itself has positive dispersion. The fraction of variance explained is called the coefficient of determination.

A partial Ordering of Dispersion

A mean-preserving spread (MPS) is a change from one probability distribution A to another probability distribution B, where B is formed by spreading out one or more portions of A's probability density function while leaving the mean (the expected value) unchanged. The concept of a mean-preserving spread provides a partial ordering of probability distributions according to their dispersions: of two probability distributions, one may be ranked as having more dispersion than the other, or alternatively neither may be ranked as having more dispersion.

Random Variable

In probability and statistics, a random variable, random quantity, aleatory variable or stochastic variable is a variable whose value is subject to variations due to chance (i.e. randomness, in a mathematical sense).A random variable can take on a set of possible different values (similarly to other mathematical variables), each with an associated probability, in contrast to other mathematical variables.

A random variable's possible values might represent the possible outcomes of a yet-to-be-performed experiment, or the possible outcomes of a past experiment whose already-existing value is uncertain (for example, due to imprecise measurements or quantum uncertainty). They may also conceptually represent either the results of an "objectively" random process (such as rolling a die) or the "subjective" randomness that results from incomplete knowledge of a quantity. The meaning of the probabilities assigned to the potential values of a random variable is not part of probability theory itself but is instead related to philosophical arguments over the interpretation of probability. The mathematics works the same regardless of the particular interpretation in use.

The mathematical function describing the possible values of a random variable and their associated probabilities is known as a probability distribution. Random variables can be discrete, that is, taking any of a specified finite or countable list of values, endowed with a probability mass function characteristic of the random variable's probability distribution; or continuous, taking any numerical value in an interval or collection of intervals, via a probability density function that is characteristic of the random variable's probability distribution; or a mixture of both types. The realizations of a random variable, that is, the results of randomly choosing values according to the variable's probability distribution function, are called random variates.

The formal mathematical treatment of random variables is a topic in probability theory. In that context, a random variable is understood as a function defined on a sample space whose outputs are numerical values.

Definition

A *random variable* $X : \Omega \to E$ is a measurable function from the set of possible outcomes Ω to some set E. The technical axiomatic definition requires Ω to be a probability space and E to be a measurable space.

Note that although X is usually a real-valued function ($E = \mathbb{R}$), it does *not* return a probability. The probabilities of different outcomes or sets of outcomes (events) are already given by the probability measure P with which Ω is equipped. Rather, X describes some numerical property that outcomes in Ω may have — e.g., the number of heads in a random collection of coin flips, or the height of a random person. The probability that X takes value ≤ 3 is the measure of the set of outcomes $\{\omega \in \Omega : X(\omega) \leq 3\}$, denoted $P(X \leq 3)$.

Standard Case

Usually $E = \mathbb{R}$. Otherwise the term *random element* is used.

When the image (or range) of X is finite or countably infinite, the random variable is called a discrete random variable and its distribution can be described by a probability mass function which assigns a probability to each value in the image of X. If the image is uncountably infinite then X is called a continuous random variable. In the special case that it is absolutely continuous, its distribution can be described by a probability density function, which assigns probabilities to intervals; in particular, each individual point must necessarily have probability zero for an absolutely continuous random variable. Not all continuous random variables are absolutely continuous, for example a mixture distribution. Such random variables cannot be described by a probability density or a probability mass function.

Any random variable can be described by its cumulative distribution function, which describes the probability that the random variable will be less than or equal to a certain value.

Extensions

The term "random variable" in statistics is traditionally limited to the real-valued case ($E = \mathbb{R}$). This ensures that it is possible to define quantities such as the expected value and variance of a random variable, its cumulative distribution function, and the moments of its distribution.

However, the definition above is valid for any measurable space E of values. Thus one can consider random elements of other sets E, such as random boolean values, categorical values, complex numbers, vectors, matrices, sequences, trees, sets, shapes, manifolds, and functions. One may then specifically refer to a *random variable of type E*, or an *E-valued random variable*.

This more general concept of a random element is particularly useful in disciplines such as graph theory, machine learning, natural language processing, and other fields in discrete mathematics and computer science, where one is often interested in modeling the random variation of non-numerical data structures. In some cases, it is nonetheless convenient to represent each element of E using one or more real numbers. In this case, a random element may optionally be represented as a vector of real-valued random variables (all defined on the same underlying probability space Ω, which allows the different random variables to covary). For example:

- A random word may be represented as a random integer that serves as an index into the vocabulary of possible words. Alternatively, it can be represented as a random indicator vector whose length equals the size of the vocabulary, where the only values of positive probability are $(1\,0\,0\,0\cdots)$, $(0\,1\,0\,0\cdots)$, $(0\,0\,1\,0\cdots)$ and the position of the 1 indicates the word.

- A random sentence of given length N may be represented as a vector of N random words.

- A random graph on N given vertices may be represented as a $N \times N$ matrix of random variables, whose values specify the adjacency matrix of the random graph.

- A random function F may be represented as a collection of random variables $F(x)$, giving the function's values at the various points x in the function's domain. The $F(x)$ are ordinary real-valued random variables provided that the function is real-valued. For example, a stochastic process is a random function of time, a random vector is a random function of some index set such as

$1, 2, \ldots n,$, and random field is a random function on any set (typically time, space, or a discrete set).

Examples

Discrete Random Variable

In an experiment a person may be chosen at random, and one random variable may be the person's height. Mathematically, the random variable is interpreted as a function which maps the person to the person's height. Associated with the random variable is a probability distribution that allows the computation of the probability that the height is in any subset of possible values, such as the probability that the height is between 180 and 190 cm, or the probability that the height is either less than 150 or more than 200 cm.

Another random variable may be the person's number of children; this is a discrete random variable with non-negative integer values. It allows the computation of probabilities for individual integer values – the probability mass function (PMF) – or for sets of values, including infinite sets. For example, the event of interest may be "an even number of children". For both finite and infinite event sets, their probabilities can be found by adding up the PMFs of the elements; that is, the probability of an even number of children is the infinite sum PMF

$$\text{PMF}(0) + \text{PMF}(2) + \text{PMF}(4) + \cdots.$$

In examples such as these, the sample space (the set of all possible persons) is often suppressed, since it is mathematically hard to describe, and the possible values of the random variables are then treated as a sample space. But when two random variables are measured on the same sample space of outcomes, such as the height and number of children being computed on the same random persons, it is easier to track their relationship if it is acknowledged that both height and number of children come from the same random person, for example so that questions of whether such random variables are correlated or not can be posed.

Coin Toss

The possible outcomes for one coin toss can be described by the sample space $\Omega = \{\text{heads, tails}\}$. We can introduce a real-valued random variable Y that models a \$1 payoff for a successful bet on heads as follows:

$$Y(\omega) = \begin{cases} 1, & \text{if } \omega = \text{heads}, \\ 0, & \text{if } \omega = \text{tails}. \end{cases}$$

If the coin is a fair coin, Y has a probability mass function f_Y given by:

$$f_Y(y) = \begin{cases} \frac{1}{2}, & \text{if } y = 1, \\ \\ \frac{1}{2}, & \text{if } y = 0, \end{cases}$$

Dice Roll

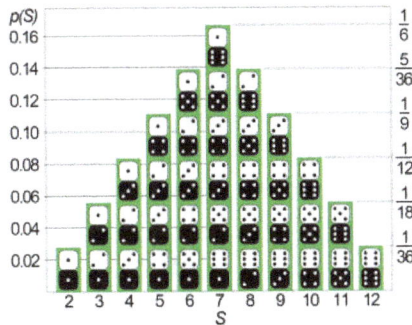

If the sample space is the set of possible numbers rolled on two dice, and the random variable of interest is the sum S of the numbers on the two dice, then S is a discrete random variable whose distribution is described by the probability mass function plotted as the height of picture columns here.

A random variable can also be used to describe the process of rolling dice and the possible outcomes. The most obvious representation for the two-dice case is to take the set of pairs of numbers n_1 and n_2 from {1, 2, 3, 4, 5, 6} (representing the numbers on the two dice) as the sample space. The total number rolled (the sum of the numbers in each pair) is then a random variable X given by the function that maps the pair to the sum:

$$X((n_1, n_2)) = n_1 + n_2$$

and (if the dice are fair) has a probability mass function f_X given by:

$$f_X(S) = \frac{\min(S-1, 13-S)}{36}, \text{ for } S \in \{2, 3, 4, 5, 6, 7, 8, 9, 10, 11, 12\}$$

Continuous Random Variable

An example of a continuous random variable would be one based on a spinner that can choose a horizontal direction. Then the values taken by the random variable are directions. We could represent these directions by North, West, East, South, Southeast, etc. However, it is commonly more convenient to map the sample space to a random variable which takes values which are real numbers. This can be done, for example, by mapping a direction to a bearing in degrees clockwise from North. The random variable then takes values which are real numbers from the interval [0, 360), with all parts of the range being "equally likely". In this case, X = the angle spun. Any real number has probability zero of being selected, but a positive probability can be assigned to any *range* of values. For example, the probability of choosing a number in [0, 180] is $\frac{1}{2}$. Instead of speaking of a probability mass function, we say that the probability *density* of X is 1/360. The probability of a subset of [0, 360) can be calculated by multiplying the measure of the set by 1/360. In general, the probability of a set for a given continuous random variable can be calculated by integrating the density over the given set.

Mixed Type

An example of a random variable of mixed type would be based on an experiment where a coin is flipped and the spinner is spun only if the result of the coin toss is heads. If the result is tails, $X = -1$; otherwise X = the value of the spinner as in the preceding example. There is a probability of $\frac{1}{2}$ that this random variable will have the value -1. Other ranges of values would have half the probabilities of the last example.

Measure-theoretic Definition

The most formal, axiomatic definition of a random variable involves measure theory. Continuous random variables are defined in terms of sets of numbers, along with functions that map such sets to probabilities. Because of various difficulties (e.g. the Banach–Tarski paradox) that arise if such sets are insufficiently constrained, it is necessary to introduce what is termed a sigma-algebra to constrain the possible sets over which probabilities can be defined. Normally, a particular such sigma-algebra is used, the Borel σ-algebra, which allows for probabilities to be defined over any sets that can be derived either directly from continuous intervals of numbers or by a finite or countably infinite number of unions and/or intersections of such intervals.

The measure-theoretic definition is as follows.

Let $(\Omega, \mathcal{F}, P)$ be a probability space and $(E, \mathcal{E})$ a measurable space. Then an $(E, \mathcal{E})$-valued random variable is a function $X: \Omega \to E$ which is $(\mathcal{F}, \mathcal{E})$-measurable. The latter means that, for every subset $B \in \mathcal{E}, its$, its preimage $X^{-1}(B) \in \mathcal{F}$ where $X^{-1}(B) = \{\omega : X(\omega) \in B\}$. This definition enables us to measure any subset $B \in \mathcal{E}in$ in the target space by looking at its preimage, which by assumption is measurable.

In more intuitive terms, Ω represents the "outcome", F represents the measurable subsets of possible outcomes, P represents the function that gives the probability of any such subset, E represents the kind of quantity the random value should take (such as real numbers), and $\mathcal{E}$ represents all the "well-behaved" (measurable) subsets of E (those for which you might want to find the probability). The random variable is then a function from any outcome to a quantity, such that the outcomes leading to any useful subset of quantities for the random variable have a well-defined probability.

When E is a topological space, then the most common choice for the σ-algebra $\mathcal{E}$

is the Borel σ-algebra $\mathcal{B}(E)$, which is the σ-algebra generated by the collection of all open sets in E. In such case the $(E, \mathcal{E})$-valued random variable is called the E-valued random variable. Moreover, when space E is the real line $\mathbb{R}$, then such a real-valued random variable is called simply the random variable.

Real-valued Random Variables

In this case the observation space is the set of real numbers. Recall, $(\Omega, \mathcal{F}, P)$ is the probability space. For real observation space, the function $X : \Omega \to \mathbb{R}$ is a real-valued random variable if

$$\{\omega : X(\omega) \leq r\} \in \mathcal{F} \qquad \forall r \in \mathbb{R}.$$

This definition is a special case of the above because the set $\{(-\infty, r] : r \in \mathbb{R}\}$ generates the Borel σ-algebra on the set of real numbers, and it suffices to check measurability on any generating set. Here we can prove measurability on this generating set by using the fact that

$$\{\omega : X(\omega) \leq r\} = X^{-1}((-\infty, r]).$$

Distribution Functions of Random Variables

If a random variable $X : \Omega \to \mathbb{R}$ defined on the probability space $(\Omega, \mathcal{F}, P)$ is given, we can ask questions like "How likely is it that the value of X is equal to 2?". This is the same as the probability of the event $\{\omega : X(\omega) = 2\}$ which is often written as $P(X = 2)$ or $p_X(2)$ for short.

Recording all these probabilities of output ranges of a real-valued random variable X yields the probability distribution of X. The probability distribution "forgets" about the particular probability space used to define X and only records the probabilities of various values of X. Such a probability distribution can always be captured by its cumulative distribution function

$$F_X(x) = P(X \leq x)$$

and sometimes also using a probability density function, p_X. In measure-theoretic terms, we use the random variable X to "push-forward" the measure P on Ω to a measure p_X on $\mathbb{R}$. The underlying probability space Ω is a technical device used to guarantee the existence of random variables, sometimes to construct them, and to define notions such as correlation and dependence or independence based on a joint distribution of two or more random variables on the same probability space. In practice, one often disposes of the space Ω altogether and just puts a measure on $\mathbb{R}$. that assigns measure 1 to the whole real line, i.e., one works with probability distributions instead of random variables.

Moments

The probability distribution of a random variable is often characterised by a small number of parameters, which also have a practical interpretation. For example, it is often enough to know what its "average value" is. This is captured by the mathematical concept of expected value of a random variable, denoted $E[X]$, and also called the first moment. In general, $E[f(X)]$ is not equal to $f(E[X])$.. Once the "average value" is known, one could then ask how far from this average value the values of X typically are, a question that is answered by the variance and standard deviation of a random variable. $E[X]$ can be viewed intuitively as an average obtained from an infinite population, the members of which are particular evaluations of X.

Mathematically, this is known as the (generalised) problem of moments: for a given class of random variables X, find a collection $\{f_i\}$ of functions such that the expectation values $E[f_i(X)]$ fully characterise the distribution of the random variable X.

Moments can only be defined for real-valued functions of random variables (or complex-valued, etc.). If the random variable is itself real-valued, then moments of the variable itself can be taken, which are equivalent to moments of the identity function $f(X) = X$ of the random variable. However, even for non-real-valued random variables, moments can be taken of real-valued functions of those variables. For example, for a categorical random variable X that can take on the nominal values "red", "blue" or "green", the real-valued function $[X = \text{green}]$ can be constructed; this uses the Iverson bracket, and has the value 1 if X has the value "green", 0 otherwise. Then, the expected value and other moments of this function can be determined.

Functions of Random Variables

A new random variable Y can be defined by applying a real Borel measurable function $g : \mathbb{R} \to \mathbb{R}$ to the outcomes of a real-valued random variable X. The cumulative distribution function of Y is

$$F_Y(y) = P(g(X) \le y).$$

If function g is invertible, i.e., g^{-1} exists, and is either increasing or decreasing, then the previous relation can be extended to obtain

$$F_Y(y) = P(g(X) \le y) = \begin{cases} P(X \le g^{-1}(y)) = F_X(g^{-1}(y)), & \text{if } g^{-1} \text{ increasing,} \\ P(X \ge g^{-1}(y)) = 1 - F_X(g^{-1}(y)), & \text{if } g^{-1} \text{ decreasing.} \end{cases}$$

and, again with the same hypotheses of invertibility of g, assuming also differentiability, we can find the relation between the probability density functions by differentiating both sides with respect to y, in order to obtain

$$f_Y(y) = f_X(g^{-1}(y)) \left| \frac{dg^{-1}(y)}{dy} \right|.$$

If there is no invertibility of g but each y admits at most a countable number of roots (i.e., a finite, or countably infinite, number of x_i such that $y = g(x_i)$) then the previous relation between the probability density functions can be generalized with

$$f_Y(y) = \sum_i f_X(g_i^{-1}(y)) \left| \frac{dg_i^{-1}(y)}{dy} \right|$$

where $x_i = g_i^{-1}(y)$.. The formulas for densities do not demand g to be increasing.

In the measure-theoretic, axiomatic approach to probability, if we have a random variable X on   and a Borel measurable function $g : \mathbb{R} \to \mathbb{R}$, then $Y = g(X)$ will also be a random variable on Ω, since the composition of measurable functions is also measurable. (However, this is not true if g is Lebesgue measurable.) The same procedure that allowed one to go from a probability space (Ω, P) to $(\mathbb{R}, dF_X)$ can be used to obtain the distribution of Y.

Example 1

Let X be a real-valued, continuous random variable and let $Y = X^2$.

$$F_Y(y) = P(X^2 \leq y).$$

If $y < 0$, then $P(X^2 \leq y) = 0$, so

$$F_Y(y) = 0 \qquad \text{if} \quad y < 0.$$

If $y \geq 0$, then

$$P(X^2 \leq y) = P(|X| \leq \sqrt{y}) = P(-\sqrt{y} \leq X \leq \sqrt{y}),$$

so

$$F_Y(y) = F_X(\sqrt{y}) - F_X(-\sqrt{y}) \qquad \text{if} \quad y \geq 0.$$

Example 2

Suppose X is a random variable with a cumulative distribution

$$F_X(x) = P(X \le x) = \frac{1}{(1+e^{-x})^\theta}$$

where $\theta > 0$ is a fixed parameter. Consider the random variable $Y = \log(1+e^{-X})$. Then,

$$F_Y(y) = P(Y \le y) = P(\log(1+e^{-X}) \le y) = P(X > -\log(e^y - 1)).$$

The last expression can be calculated in terms of the cumulative distribution of X, so

$$F_Y(y) = 1 - F_X(-\log(e^y - 1))$$

$$= 1 - \frac{1}{(1+e^{\log(e^y-1)})^\theta}$$

$$= 1 - \frac{1}{(1+e^y-1)^\theta}$$

$$= 1 - e^{-y\theta}.$$

which is the cdf of an exponential distribution.

Example 3

Suppose X is a random variable with a standard normal distribution, whose density is

$$f_X(x) = \frac{1}{\sqrt{2\pi}} e^{-x^2/2}.$$

Consider the random variable $Y = X^2$. We can find the density using the above formula for a change of variables:

$$f_Y(y) = \sum_i f_X(g_i^{-1}(y)) \left| \frac{dg_i^{-1}(y)}{dy} \right|.$$

In this case the change is not monotonic, because every value of Y has two corresponding values of X (one positive and negative). However, because of symmetry, both halves will transform identically, i.e.,

$$f_Y(y) = 2 f_X(g^{-1}(y)) \left| \frac{dg^{-1}(y)}{dy} \right|.$$

The inverse transformation is

$$x = g^{-1}(y) = \sqrt{y}$$

and its derivative is

$$\frac{dg^{-1}(y)}{dy} = \frac{1}{2\sqrt{y}}.$$

Then,

$$f_Y(y) = 2\frac{1}{\sqrt{2\pi}}e^{-y/2}\frac{1}{2\sqrt{y}} = \frac{1}{\sqrt{2\pi y}}e^{-y/2}.$$

This is a chi-squared distribution with one degree of freedom.

Equivalence of Random Variables

There are several different senses in which random variables can be considered to be equivalent. Two random variables can be equal, equal almost surely, or equal in distribution.

In increasing order of strength, the precise definition of these notions of equivalence is given below.

Equality in Distribution

If the sample space is a subset of the real line, random variables X and Y are *equal in distribution* (denoted $X \overset{d}{=} Y$) if they have the same distribution functions:

$$P(X \leq x) = P(Y \leq x) \quad \text{for all} \quad x.$$

Two random variables having equal moment generating functions have the same distribution. This provides, for example, a useful method of checking equality of certain functions of i.i.d. random variables. However, the moment generating function exists only for distributions that have a defined Laplace transform.

Almost Sure Equality

Two random variables X and Y are *equal almost surely* if, and only if, the probability that they are different is zero:

$$P(X \neq Y) = 0.$$

For all practical purposes in probability theory, this notion of equivalence is as strong as actual equality. It is associated to the following distance:

$$d_\infty(X,Y) = \operatorname*{ess\,sup}_\omega |X(\omega) - Y(\omega)|,$$

where "ess sup" represents the essential supremum in the sense of measure theory.

Equality

Finally, the two random variables X and Y are *equal* if they are equal as functions on their measurable space:

$$X(\omega) = Y(\omega) \qquad \text{for all } \omega.$$

Convergence

A significant theme in mathematical statistics consists of obtaining convergence results for certain sequences of random variables; for instance the law of large numbers and the central limit theorem.

There are various senses in which a sequence (X_n) of random variables can converge to a random variable X. These are explained in the article on convergence of random variables.

Errors and Residuals

In statistics and optimization, errors and residuals are two closely related and easily confused measures of the deviation of an observed value of an element of a statistical sample from its "theoretical value". The error (or disturbance) of an observed value is the deviation of the observed value from the (unobservable) *true* value of a quantity of interest (for example, a population mean), and the residual of an observed value is the difference between the observed value and the *estimated* value of the quantity of interest (for example, a sample mean). The distinction is most important in regression analysis, where the concepts are sometimes called the regression errors and regression residuals and where they lead to the concept of studentized residuals.

Introduction

Suppose there is a series of observations from a univariate distribution and we want to estimate the mean of that distribution (the so-called location model). In this case, the errors are the deviations of the observations from the population mean, while the residuals are the deviations of the observations from the sample mean.

A statistical error (or disturbance) is the amount by which an observation differs from its expected value, the latter being based on the whole population from which the statistical unit was chosen randomly. For example, if the mean height in a population of 21-year-old men is 1.75 meters, and one randomly chosen man is 1.80 meters tall, then the "error" is 0.05 meters; if the randomly chosen man is 1.70 meters tall, then the "error" is −0.05 meters. The expected value, being the mean of the entire population, is typically unobservable, and hence the statistical error cannot be observed either.

A residual (or fitting deviation), on the other hand, is an observable *estimate* of the unobservable statistical error. Consider the previous example with men's heights and suppose we have a random sample of *n* people. The *sample mean* could serve as a good estimator of the *population* mean. Then we have:

- The difference between the height of each man in the sample and the unobservable *population* mean is a *statistical error*, whereas

- The difference between the height of each man in the sample and the observable *sample* mean is a *residual*.

Note that the sum of the residuals within a random sample is necessarily zero, and thus the residuals are necessarily *not independent*. The statistical errors on the other hand are independent, and their sum within the random sample is almost surely not zero.

One can standardize statistical errors (especially of a normal distribution) in a z-score (or "standard score"), and standardize residuals in a *t*-statistic, or more generally studentized residuals.

In Univariate Distributions

If we assume a normally distributed population with mean μ and standard deviation σ, and choose individuals independently, then we have

$$X_1,...,X_n \sim N(\mu,\sigma^2)$$

and the sample mean

$$\bar{X} = \frac{X_1 + \cdots + X_n}{n}$$

is a random variable distributed thus:

$$\bar{X} \sim N\left(\mu,\frac{\sigma^2}{n}\right).$$

The *statistical errors* are then

$$e_i = X_i - \mu,$$

whereas the *residuals* are

$$r_i = X_i - \bar{X}.$$

The sum of squares of the statistical errors, divided by σ^2, has a chi-squared distribution with n degrees of freedom:

$$\frac{1}{\sigma^2} \sum_{i=1}^{n} e_i^2 \sim \chi_n^2.$$

This quantity, however, is not observable. The sum of squares of the residuals, on the other hand, is observable. The quotient of that sum by σ^2 has a chi-squared distribution with only $n-1$ degrees of freedom:

$$\frac{1}{\sigma^2} \sum_{i=1}^{n} r_i^2 \sim \chi_{n-1}^2.$$

This difference between n and $n-1$ degrees of freedom results in Bessel's correction for the estimation of sample variance of a population with unknown mean and unknown variance, though if the mean is known, no correction is necessary.

Remark

It is remarkable that the sum of squares of the residuals and the sample mean can be shown to be independent of each other, using, e.g. Basu's theorem. That fact, and the normal and chi-squared distributions given above, form the basis of calculations involving the quotient

$$\frac{\bar{X}_n - \mu}{S_n / \sqrt{n}},$$

which is generally called t-statistic.

The probability distributions of the numerator and the denominator separately depend on the value of the unobservable population standard deviation σ, but σ appears in both the numerator and the denominator and cancels. That is fortunate because it means that even though we do not know σ, we know the probability distribution of this quotient: it has a Student's t-distribution with $n-1$ degrees of freedom. We can therefore use this quotient to find a confidence interval for μ.

Regressions

In regression analysis, the distinction between *errors* and *residuals* is subtle and important, and leads to the concept of studentized residuals. Given an unobservable function that relates the independent variable to the dependent variable – say, a line – the deviations of the dependent variable observations from this function are the unobservable errors. If one runs a regression on some data, then the deviations of the dependent variable observations from the *fitted* function are the residuals.

However, a terminological difference arises in the expression mean squared error (MSE). The mean squared error of a regression is a number computed from the sum of squares of the computed *residuals*, and not of the unobservable *errors*. If that sum of squares is divided by n, the number of observations, the result is the mean of the squared residuals. Since this is a biased estimate of the variance of the unobserved errors, the bias is removed by multiplying the mean of the squared residuals by n / df where df is the number of degrees of freedom (n minus the number of parameters being estimated). This latter formula serves as an unbiased estimate of the variance of the unobserved errors, and is called the mean squared error.

Another method to calculate the mean square of error when analyzing the variance of linear regression using a technique like that used in ANOVA (they are the same because ANOVA is a type of regression), the sum of squares of the residuals (aka sum of squares of the error) is divided by the degrees of freedom (where the degrees of freedom equals $n-p-1$, where p is the number of 'parameters' or predictors used in the model (i.e. the number of variables in the regression equation). One can then also calculate the mean square of the model by dividing the sum of squares of the model minus the degrees of freedom, which is just the number of parameters. Then the F value can be calculated by divided MS(model) by MS(error), and we can then determine significance (which is why you want the mean squares to begin with.).

However, because of the behavior of the process of regression, the *distributions* of residuals at different data points (of the input variable) may vary *even if* the errors themselves are identically distributed. Concretely, in a linear regression where the errors are identically distributed, the variability of residuals of inputs in the middle of the domain will be *higher* than the variability of residuals at the ends of the domain: linear regressions fit endpoints better than the middle. This is also reflected in the influence functions of various data points on the regression coefficients: endpoints have more influence.

Thus to compare residuals at different inputs, one needs to adjust the residuals by the expected variability of *residuals,* which is called studentizing. This is particularly important in the case of detecting outliers: a large residual may be expected in the middle of the domain, but considered an outlier at the end of the domain.

Other uses of the Word "Error" in Statistics

The use of the term "error" as discussed in the sections above is in the sense of a deviation of a value from a hypothetical unobserved value. At least two other uses also occur in statistics, both referring to observable prediction errors:

Mean square error or mean squared error (abbreviated MSE) and root mean square error (RMSE) refer to the amount by which the values predicted by an estimator differ from the quantities being estimated (typically outside the sample from which the model was estimated).

Sum of squared errors, typically abbreviated SSE or SS_e, refers to the residual sum of squares (the sum of squared residuals) of a regression; this is the sum of the squares of the deviations of the actual values from the predicted values, within the sample used for estimation. Likewise, the sum of absolute errors (SAE) refers to the sum of the absolute values of the residuals, which is minimized in the least absolute deviations approach to regression.

Types of Errors and Residuals

Mean Squared Error

In statistics, the mean squared error (MSE) or mean squared deviation (MSD) of an estimator (of a procedure for estimating an unobserved quantity) measures the average of the squares of the errors or deviations—that is, the difference between the estimator and what is estimated. MSE is a risk function, corresponding to the expected value of the squared error loss or quadratic loss. The difference occurs because of randomness or because the estimator doesn't account for information that could produce a more accurate estimate.

The MSE is a measure of the quality of an estimator—it is always non-negative, and values closer to zero are better.

The MSE is the second moment (about the origin) of the error, and thus incorporates both the variance of the estimator and its bias. For an unbiased estimator, the MSE is the variance of the estimator. Like the variance, MSE has the same units of measurement as the square of the quantity being estimated. In an analogy to standard deviation, taking the square root of MSE yields the root-mean-square error or root-mean-square deviation (RMSE or RMSD), which has the same units as the quantity being estimated; for an unbiased estimator, the RMSE is the square root of the variance, known as the standard deviation.

Definition and Basic Properties

The MSE assesses the quality of an estimator (i.e., a mathematical function mapping a sample of data to a parameter of the population from which the data is sampled) or a predictor (i.e., a function mapping arbitrary inputs to a sample of values of some random variable). Definition of an MSE differs according to whether one is describing an estimator or a predictor.

Predictor

If $\hat{Y}$ is a vector of n predictions, and Y is the vector of observed values corresponding to the inputs to the function which generated the predictions, then the MSE of the predictor can be estimated by

$$\mathrm{MSE} = \frac{1}{n}\sum_{i=1}^{n}(\hat{Y}_i - Y_i)^2$$

I.e., the MSE is the *mean* $\left(\frac{1}{n}\sum_{i=1}^{n}\right)$ of the *square of the errors* ($(\hat{Y}_i - Y_i)^2$). This is an easily computable quantity for a particular sample (and hence is sample-dependent).

Estimator

The MSE of an estimator $\hat{\theta}$ with respect to an unknown parameter θ is defined as

$$\mathrm{MSE}(\hat{\theta}) = \mathbb{E}\left[(\hat{\theta} - \theta)^2\right].$$

This definition depends on the unknown parameter, and the MSE in this sense is a property of an estimator. Since an MSE is an expectation, it is not technically a random variable. That being said, the MSE could be a function of unknown parameters, in which case any *estimator* of the MSE based on estimates of these parameters would be a function of the data and thus a random variable. If the estimator is derived from a sample statistic and is used to estimate some population statistic, then the expectation is with respect to the sampling distribution of the sample statistic.

The MSE can be written as the sum of the variance of the estimator and the squared bias of the estimator, providing a useful way to calculate the MSE and implying that in the case of unbiased estimators, the MSE and variance are equivalent.

$$\mathrm{MSE}(\hat{\theta}) = \mathrm{Var}(\hat{\theta}) + \mathrm{Bias}(\hat{\theta}, \theta)^2.$$

Proof of Variance and Bias Relationship

$$
\begin{aligned}
\mathrm{MSE}(\hat{\theta}) &= \mathbb{E}\left[(\hat{\theta} - \theta)^2\right] \\
&= \mathbb{E}\left[\left(\hat{\theta} - \mathbb{E}[\hat{\theta}] + \mathbb{E}[\hat{\theta}] - \theta\right)^2\right] \\
&= \mathbb{E}\left[\left(\hat{\theta} - \mathbb{E}[\hat{\theta}]\right)^2 + 2\left(\hat{\theta} - \mathbb{E}[\hat{\theta}]\right)\left(\mathbb{E}[\hat{\theta}] - \theta\right) + \left(\mathbb{E}[\hat{\theta}] - \theta\right)^2\right] \\
&= \mathbb{E}\left[\left(\hat{\theta} - \mathbb{E}[\hat{\theta}]\right)^2\right] + \mathbb{E}\left[2\left(\hat{\theta} - \mathbb{E}[\hat{\theta}]\right)\left(\mathbb{E}[\hat{\theta}] - \theta\right)\right] + \mathbb{E}\left[\left(\mathbb{E}[\hat{\theta}] - \theta\right)^2\right] \\
&= \mathbb{E}\left[\left(\hat{\theta} - \mathbb{E}[\hat{\theta}]\right)^2\right] + 2\left(\mathbb{E}[\hat{\theta}] - \theta\right)\mathbb{E}\left[\hat{\theta} - \mathbb{E}[\hat{\theta}]\right] + \left(\mathbb{E}[\hat{\theta}] - \theta\right)^2 \qquad \mathbb{E}[\hat{\theta}] - \theta = \mathrm{const.} \\
&= \mathbb{E}\left[\left(\hat{\theta} - \mathbb{E}[\hat{\theta}]\right)^2\right] + 2\left(\mathbb{E}[\hat{\theta}] - \theta\right)\left(\mathbb{E}[\hat{\theta}] - \mathbb{E}[\hat{\theta}]\right) + \left(\mathbb{E}[\hat{\theta}] - \theta\right)^2 \qquad \mathbb{E}[\hat{\theta}] = \mathrm{const.} \\
&= \mathbb{E}\left[\left(\hat{\theta} - \mathbb{E}[\hat{\theta}]\right)^2\right] + \left(\mathbb{E}[\hat{\theta}] - \theta\right)^2 \\
&= \mathrm{Var}(\hat{\theta}) + \mathrm{Bias}(\hat{\theta}, \theta)^2
\end{aligned}
$$

Regression

In regression analysis, the term *mean squared error* is sometimes used to refer to the unbiased estimate of error variance: the residual sum of squares divided by the number of degrees of freedom. This definition for a known, computed quantity differs from the above definition for the computed MSE of a predictor in that a different denominator is used. The denominator is the sample size reduced by the number of model parameters estimated from the same data, $(n-p)$ for p regressors or $(n-p-1)$ if an intercept is used. For more details. Note that, although the MSE is not an unbiased estimator of the error variance, it is consistent, given the consistency of the predictor.

Also in regression analysis, "mean squared error", often referred to as mean squared prediction error or "out-of-sample mean squared error", can refer to the mean value of the squared deviations of the predictions from the true values, over an out-of-sample test space, generated by a model estimated over a particular sample space. This also is a known, computed quantity, and it varies by sample and by out-of-sample test space.

Examples

Mean

Suppose we have a random sample of size n from a population, $X_1, \ldots, X_n$. Suppose the sample units were chosen with replacement. That is, the n units are selected one at a time, and previously selected units are still eligible for selection for all n draws. The usual estimator for the mean is the sample average

$$\bar{X} = \frac{1}{n} \sum_{i=1}^{n} X_i$$

which has an expected value equal to the true mean μ (so it is unbiased) and a mean square error of

$$\mathrm{MSE}\left(\bar{X}\right) = \mathrm{E}\left[\left(\bar{X} - \mu\right)^2\right] = \left(\frac{\sigma}{\sqrt{n}}\right)^2 = \frac{\sigma^2}{n}$$

where σ^2 is the population variance.

For a Gaussian distribution this is the best unbiased estimator (that is, it has the lowest MSE among all unbiased estimators), but not, say, for a uniform distribution.

Variance

The usual estimator for the variance is the *corrected sample variance:*

$$S_{n-1}^2 = \frac{1}{n-1}\sum_{i=1}^{n}\left(X_i - \bar{X}\right)^2 = \frac{1}{n-1}\left(\sum_{i=1}^{n}X_i^2 - n\bar{X}^2\right).$$

This is unbiased (its expected value is σ^2), hence also called the *unbiased sample variance,* and its MSE is

$$\text{MSE}(S_{n-1}^2) = \frac{1}{n}\left(\mu_4 - \frac{n-3}{n-1}\sigma^4\right) = \frac{1}{n}\left(\gamma_2 + \frac{2n}{n-1}\right)\sigma^4,$$

where μ_4 is the fourth central moment of the distribution or population and $\gamma_2 = \mu_4/\sigma^4 - 3$ is the excess kurtosis.

However, one can use other estimators for σ^2 which are proportional to S_{n-1}^2, and an appropriate choice can always give a lower mean square error. If we define

$$S_a^2 = \frac{n-1}{a}S_{n-1}^2 = \frac{1}{a}\sum_{i=1}^{n}\left(X_i - \bar{X}\right)^2$$

then we calculate:

$$\text{MSE}(S_a^2) = \mathbb{E}\left[\left(\frac{n-1}{a}S_{n-1}^2 - \sigma^2\right)^2\right]$$

$$= \mathbb{E}\left[\frac{(n-1)^2}{a^2}S_{n-1}^4 - 2\left(\frac{n-1}{a}S_{n-1}^2\right)\sigma^2 + \sigma^4\right]$$

$$= \frac{(n-1)^2}{a^2}\mathbb{E}[S_{n-1}^4] - 2\left(\frac{n-1}{a}\right)\mathbb{E}[S_{n-1}^2]\sigma^2 + \sigma^4$$

$$= \frac{(n-1)^2}{a^2}\mathbb{E}[S_{n-1}^4] - 2\left(\frac{n-1}{a}\right)\sigma^4 + \sigma^4 \qquad \mathbb{E}[S_{n-1}^2] = \sigma^2$$

$$= \frac{(n-1)^2}{a^2}\left(\frac{\gamma_2}{n} + \frac{n+1}{n-1}\right)\sigma^4 - 2\left(\frac{n-1}{a}\right)\sigma^4 + \sigma^4 \qquad \mathbb{E}[S_{n-1}^4] = \text{MSE}(S_{n-1}^2) + \sigma^4$$

$$= \frac{n-1}{na^2}\left((n-1)\gamma_2 + n^2 + n\right)\sigma^4 - 2\left(\frac{n-1}{a}\right)\sigma^4 + \sigma^4$$

This is minimized when

$$a = \frac{(n-1)\gamma_2 + n^2 + n}{n} = n+1+\frac{n-1}{n}\gamma_2.$$

For a Gaussian distribution, where $\gamma_2 = 0$, , this means the MSE is minimized when dividing the sum by $a = n+1$.. The minimum excess kurtosis is $\gamma_2 = -2$, , which is achieved by a Bernoulli distribution with $p = 1/2$ (a coin flip), and the MSE is minimized for $a = n - 1 + \frac{2}{n}$. So no matter what the kurtosis, we get a "better" estimate (in the sense of having a lower MSE) by scaling down the unbiased estimator a little bit; this is a simple example of a shrinkage estimator: one "shrinks" the estimator towards zero (scales down the unbiased estimator).

Further, while the corrected sample variance is the best unbiased estimator (minimum mean square error among unbiased estimators) of variance for Gaussian distributions, if the distribution is not Gaussian then even among unbiased estimators, the best unbiased estimator of the variance may not be S_{n-1}^2.

Gaussian Distribution

The following table gives several estimators of the true parameters of the population, μ and σ^2, for the Gaussian case.

True value	Estimator	Mean squared error
$\theta = \mu$	$\hat{\theta} =$ the unbiased estimator of the population mean, $\bar{X} = \frac{1}{n}\sum_{i=1}^{n}(X_i)$	$\mathrm{MSE}(\bar{X}) = \mathrm{E}((\bar{X} - \mu)^2) = \left(\frac{\sigma}{\sqrt{n}}\right)^2$
$\theta = \sigma^2$	$\hat{\theta} =$ the unbiased estimator of the population variance, $S_{n-1}^2 = \frac{1}{n-1}\sum_{i=1}^{n}\left(X_i - \bar{X}\right)^2$	$\mathrm{MSE}(S_{n-1}^2) = \mathrm{E}((S_{n-1}^2 - \sigma^2)^2) = \frac{2}{n-1}\sigma^4$
$\theta = \sigma^2$	$\hat{\theta} =$ the biased estimator of the population variance, $S_n^2 = \frac{1}{n}\sum_{i=1}^{n}\left(X_i - \bar{X}\right)^2$	$\mathrm{MSE}(S_n^2) = \mathrm{E}((S_n^2 - \sigma^2)^2) = \frac{2n-1}{n^2}\sigma^4$
$\theta = \sigma^2$	$\hat{\theta} =$ the biased estimator of the population variance, $S_{n+1}^2 = \frac{1}{n+1}\sum_{i=1}^{n}\left(X_i - \bar{X}\right)^2$	$\mathrm{MSE}(S_{n+1}^2) = \mathrm{E}((S_{n+1}^2 - \sigma^2)^2) = \frac{2}{n+1}\sigma^4$

Note that:

1. The MSEs shown for the variance estimators assume $X_i \sim N(\mu, \sigma^2)$ i.i.d. so that $\frac{(n-1)S_{n-1}^2}{\sigma^2} \sim \chi_{n-1}^2$. The result for S_{n-1}^2 follows easily from the χ_{n-1}^2 variance that is $2n - 2$.

2. Unbiased estimators may not produce estimates with the smallest total variation (as measured by MSE): the MSE S_{n-1}^2 of is larger than that of S_{n+1}^2 or S_n^2.

3. Estimators with the smallest total variation may produce biased estimates: S_{n+1}^2 typically underestimates σ^2 by $\frac{2}{n}\sigma^2$

Interpretation

An MSE of zero, meaning that the estimator $\hat{\theta}$ predicts observations of the parameter θ with perfect accuracy, is the ideal, but is typically not possible.

Values of MSE may be used for comparative purposes. Two or more statistical models may be compared using their MSEs as a measure of how well they explain a given set of observations: An unbiased estimator (estimated from a statistical model) with the smallest variance among all unbiased estimators is the [best unbiased estimator]] or MVUE (Minimum Variance Unbiased Estimator).

Both linear regression techniques such as analysis of variance estimate the MSE as part of the analysis and use the estimated MSE to determine the statistical significance of the factors or predictors under study. The goal of experimental design is to construct experiments in such a way that when the observations are analyzed, the MSE is close to zero relative to the magnitude of at least one of the estimated treatment effects.

MSE is also used in several stepwise regression techniques as part of the determination as to how many predictors from a candidate set to include in a model for a given set of observations.

Applications

- Minimizing MSE is a key criterion in selecting estimators. Among unbiased estimators, minimizing the MSE is equivalent to minimizing the variance, and the estimator that does this is the minimum variance unbiased estimator. However, a biased estimator may have lower MSE.

- In statistical modelling the MSE, representing the difference between the actual observations and the observation values predicted by the model, is used to determine the extent to which the model fits the data and whether the removal or some explanatory variables, simplifying the model, is possible without significantly harming the model's predictive ability.

Loss Function

Squared error loss is one of the most widely used loss functions in statistics, though its widespread use stems more from mathematical convenience than considerations of actual loss in applications. Carl Friedrich Gauss, who introduced the use of mean squared error, was aware of its arbitrariness and was in agreement with objections to it on these grounds. The mathematical benefits of mean squared error are particularly evident in its use at analyzing the performance of linear regression, as it allows one to partition the variation in a dataset into variation explained by the model and variation explained by randomness.

Criticism

The use of mean squared error without question has been criticized by the decision theorist James Berger. Mean squared error is the negative of the expected value of one specific utility function, the quadratic utility function, which may not be the appropriate utility function to use under a given set of circumstances. There are, however, some scenarios where mean squared error can serve as a good approximation to a loss function occurring naturally in an application.

Like variance, mean squared error has the disadvantage of heavily weighting outliers. This is a result of the squaring of each term, which effectively weights large errors more heavily than small ones. This property, undesirable in many applications, has led researchers to use alternatives such as the mean absolute error, or those based on the median.

Root-mean-square Deviation

The root-mean-square deviation (RMSD) or root-mean-square error (RMSE) is a frequently used measure of the differences between values (sample and population values) predicted by a model or an estimator and the values actually observed. The RMSD represents the sample standard deviation of the differences between predicted values and observed values. These individual differences are called residuals when the calculations are performed over the data sample that was used for estimation, and are called *prediction errors* when computed out-of-sample. The RMSD serves to aggregate the magnitudes of the errors in predictions for various times into a single measure of predictive power. RMSD is a good measure of accuracy, but only to compare forecasting errors of different models for a particular variable and not between variables, as it is scale-dependent.

Formula

The RMSD of an estimator $\hat{\theta}$ with respect to an estimated parameter θ is defined as the square root of the mean square error:

$$\text{RMSD}(\hat{\theta}) = \sqrt{\text{MSE}(\hat{\theta})} = \sqrt{\text{E}((\hat{\theta} - \theta)^2)}.$$

For an unbiased estimator, the RMSD is the square root of the variance, known as the standard deviation.

The RMSD of predicted values $\hat{y}_t$ for times t of a regression's dependent variable y_t is computed for n different predictions as the square root of the mean of the squares of the deviations:

$$\text{RMSD} = \sqrt{\frac{\sum_{t=1}^{n} (\hat{y}_t - y_t)^2}{n}}.$$

In some disciplines, the RMSD is used to compare differences between two things that may vary, neither of which is accepted as the "standard". For example, when measuring the average difference between two time series $x_{1,t}$ and $x_{2,t}$, the formula becomes

$$RMSD = \sqrt{\frac{\sum_{t=1}^{n}(x_{1,t} - x_{2,t})^2}{n}}.$$

Normalized root-mean-square Deviation

Normalizing the RMSD facilitates the comparison between datasets or models with different scales. Though there is no consistent means of normalization in the literature, common choices are the mean or the range (defined as the maximum value minus the minimum value) of the measured data:

$$NRMSD = \frac{RMSD}{y_{max} - y_{min}} \text{ or } NRMSD = \frac{RMSD}{\bar{y}}.$$

This value is commonly referred to as the normalized root-mean-square deviation or error (NRMSD or NRMSE), and often expressed as a percentage, where lower values indicate less residual variance. In many cases, especially for smaller samples, the sample range is likely to be affected by the size of sample which would hamper comparisons.

When normalising by the mean value of the measurements, the term coefficient of variation of the RMSD, CV(RMSD) may be used to avoid ambiguity. This is analogous to the coefficient of variation with the RMSD taking the place of the standard deviation.

$$CV(RMSD) = \frac{RMSD}{\bar{y}}$$

Applications

- In meteorology, to see how effectively a mathematical model predicts the behavior of the atmosphere.

- In bioinformatics, the RMSD is the measure of the average distance between the atoms of superimposed proteins.

- In structure based drug design, the RMSD is a measure of the difference between a crystal conformation of the ligand conformation and a docking prediction.

- In economics, the RMSD is used to determine whether an economic model fits economic indicators. Some experts have argued that RMSD is less reliable than Relative Absolute Error.

- In experimental psychology, the RMSD is used to assess how well mathematical or computational models of behavior explain the empirically observed behavior.

- In GIS, the RMSD is one measure used to assess the accuracy of spatial analysis and remote sensing.

- In hydrogeology, RMSD and NRMSD are used to evaluate the calibration of a groundwater model.

- In imaging science, the RMSD is part of the peak signal-to-noise ratio, a measure used to assess how well a method to reconstruct an image performs relative to the original image.

- In computational neuroscience, the RMSD is used to assess how well a system learns a given model.

- In Protein nuclear magnetic resonance spectroscopy, the RMSD is used as a measure to estimate the quality of the obtained bundle of structures.

- Submissions for the Netflix Prize were judged using the RMSD from the test dataset's undisclosed "true" values.

- In simulation of energy consumption of buildings, the RMSE and CV(RMSE) are used to calibrate models to measured building performance.

- In X-ray crystallography, RMSD (and RMSZ) is used to measure the deviation of the molecular internal coordinates deviate from the restraints library values.

- Errors and residuals in statistics

Least Absolute Deviations

Least absolute deviations (LAD), also known as least absolute errors (LAE), least absolute value (LAV), least absolute residual (LAR), sum of absolute deviations, or the L_1 norm condition, is a statistical optimality criterion and the statistical optimization technique that relies on it. Similar to the popular least squares technique, it attempts to find a function which closely approximates a set of data. In the simple case of a set of (x,y) data, the approximation function is a simple "trend line" in two-dimensional Cartesian coordinates. The method minimizes the sum of absolute errors (SAE) (the sum of the absolute values of the vertical "residuals" between points generated by the function and corresponding points in the data). The least absolute deviations estimate also arises as the maximum likelihood estimate if the errors have a Laplace distribution.

Formulation of the Problem

Suppose that the data set consists of the points (x_i, y_i) with $i = 1, 2, ..., n$. We want to find a function f such that $f(x_i) \approx y_i$.

To attain this goal, we suppose that the function f is of a particular form containing some parameters which need to be determined. For instance, the simplest form would be linear: $f(x) = bx + c$, where b and c are parameters whose values are not known but which we would like to estimate. Less simply, suppose that $f(x)$ is quadratic, meaning that $f(x) = ax^2 + bx + c$, where a, b and c are not yet known. (More generally, there could be not just one explanator x, but rather multiple explanators, all appearing as arguments of the function f.)

We now seek estimated values of the unknown parameters that minimize the sum of the absolute values of the residuals:

$$S = \sum_{i=1}^{n} | y_i - f(x_i) |.$$

Contrasting Least Squares with Least Absolute Deviations

The following is a table contrasting some properties of the method of least absolute deviations with those of the method of least squares (for non-singular problems).

Least squares regression	Least absolute deviations regression
Not very robust	Robust
Stable solution	Unstable solution
Always one solution	Possibly multiple solutions

The method of least absolute deviations finds applications in many areas, due to its robustness compared to the least squares method. Least absolute deviations is robust in that it is resistant to outliers in the data. LAD gives equal emphasis to all observations, in contrast to OLS which, by squaring the residuals, gives more weight to large residuals, that is, outliers in which predicted values are far from actual observations. This may be helpful in studies where outliers do not need to be given greater weight than other observations. If it is important to give greater weight to outliers, the method of least squares is a better choice.

Other Properties

There exist other unique properties of the least absolute deviations line. In the case of a set of (x, y) data, the least absolute deviations line will always pass through at least two of the data points, unless there are multiple solutions. If multiple solutions exist, then the region of valid least absolute deviations solutions will be bounded by at least two lines, each of which passes through at least two data points. More generally, if there are k regressors (including the constant), then at least one optimal regression surface will pass through k of the data points.

This "latching" of the line to the data points can help to understand the "instability" property: if the line always latches to at least two points, then the line will jump between different sets of points as the data points are altered. The "latching" also helps to understand the "robustness" property: if there exists an outlier, and a least absolute deviations line must latch onto two data points, the outlier will most likely not be one of those two points because that will not minimize the sum of absolute deviations in most cases.

One known case in which multiple solutions exist is a set of points symmetric about a horizontal line, as shown in Figure A below.

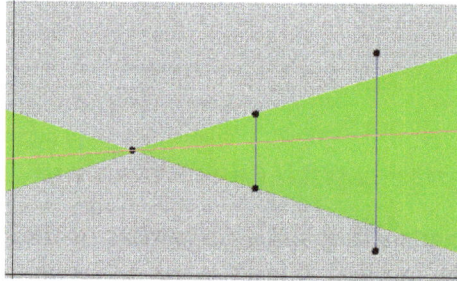

Figure A: A set of data points with reflection symmetry and multiple least absolute deviations solutions. The "solution area" is shown in green. The vertical blue lines represent the absolute errors from the pink line to each data point. The pink line is one of infinitely many solutions within the green area.

To understand why there are multiple solutions in the case shown in Figure A, consider the pink line in the green region. Its sum of absolute errors is some value S. If one were to tilt the line upward slightly, while still keeping it within the green region, the sum of errors would still be S. It would not change because the distance from each point to the line grows on one side of the line, while the distance to each point on the opposite side of the line diminishes by exactly the same amount. Thus the sum of absolute errors remains the same. Also, since one can tilt the line in infinitely small increments, this also shows that if there is more than one solution, there are infinitely many solutions.

Variations, Extensions, Specializations

The least absolute deviation problem may be extended to include multiple explanators, constraints and regularization, e.g., a linear model with linear constraints:

$$\text{minimize } S(\beta, b) = \sum_i |x_i'\beta + b - y_i|$$

$$\text{subject to, e.g., } x_1'\beta + b - y_1 \leq k$$

where β is a column vector of coefficients to be estimated, b is an intercept to be estimated, x_i is a column vector of the ith observations on the various explanators, y_i is the ith observation on the dependent variable, and k is a known constant.

Regularization with LASSO may also be combined with LAD.

Solving Methods

Though the idea of least absolute deviations regression is just as straightforward as that of least squares regression, the least absolute deviations line is not as simple to compute efficiently. Unlike least squares regression, least absolute deviations regression does not have an analytical solving method. Therefore, an iterative approach is required. The following is an enumeration of some least absolute deviations solving methods.

- Simplex-based methods (such as the Barrodale-Roberts algorithm)

  - o   Because the problem is a linear program, any of the many linear programming techniques (including the simplex method as well as others) can be applied.

- Iteratively re-weighted least squares

- Wesolowsky's direct descent method

- Li-Arce's maximum likelihood approach

- Check all combinations of point-to-point lines for minimum sum of errors

Simplex-based methods are the "preferred" way to solve the least absolute deviations problem. A Simplex method is a method for solving a problem in linear programming. The most popular algorithm is the Barrodale-Roberts modified Simplex algorithm. The algorithms for IRLS, Wesolowsky's Method, and Li's Method can be found in Appendix A of among other methods. Checking all combinations of lines traversing any two (x,y) data points is another method of finding the least absolute deviations line. Since it is known that at least one least absolute deviations line traverses at least two data points, this method will find a line by comparing the SAE (Smallest Absolute Error over data points) of each line, and choosing the line with the smallest SAE. In addition, if multiple lines have the same, smallest SAE, then the lines outline the region of multiple solutions. Though simple, this final method is inefficient for large sets of data.

Solving using Linear Programming

The problem can be solved using any linear programming technique on the following problem specification. We wish to

$$\text{Minimize} \sum_{i=1}^{n} |y_i - a_0 - a_1 x_{i1} - a_2 x_{i2} - \cdots - a_k x_{ik}|$$

with respect to the choice of the values of the parameters $a_0, \ldots, a_k$, where y_i is the value of the ith observation of the dependent variable, and x_{ij} is the value of the ith observation of the jth independent variable ($j = 1,...,k$). We rewrite this problem in terms of artificial variables u_i as

$$\text{Minimize} \sum_{i=1}^{n} u_i$$

with respect to

$$a_0,...,a_k \text{ and } u_1,...,u_n$$

subject to

$$u_i \geq y_i - a_0 - a_1 x_{i1} - a_2 x_{i2} - \cdots - a_k x_{ik} \qquad \text{for } i = 1,...,n$$

$$u_i \geq -[y_i - a_0 - a_1 x_{i1} - a_2 x_{i2} - \cdots - a_k x_{ik}] \quad \text{for } i = 1,...,n.$$

These constraints have the effect of forcing each u_i to equal $| y_i - a_0 - a_1 x_{i1} - a_2 x_{i2} - \cdots - a_k x_{ik} |$ upon being minimized, so the objective function is equivalent to the original objective function. Since this version of the problem statement does not contain the absolute value operator, it is in a format that can be solved with any linear programming package.

Least Squares

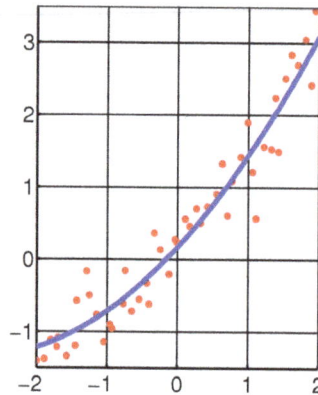

The result of fitting a set of data points with a quadratic function

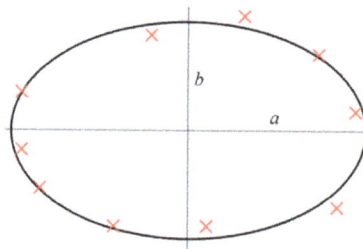

Conic fitting a set of points using least-squares approximation

The method of least squares is a standard approach in regression analysis to the approximate solution of overdetermined systems, i.e., sets of equations in which there are more equations than unknowns. "Least squares" means that the overall solution minimizes the sum of the squares of the errors made in the results of every single equation.

The most important application is in data fitting. The best fit in the least-squares sense minimizes the sum of squared residuals, a residual being the difference between an observed value and the fitted value provided by a model. When the problem has substantial uncertainties in the independent variable (the x variable), then simple regression and least squares methods have problems; in such cases, the methodology required for fitting errors-in-variables models may be considered instead of that for least squares.

Least squares problems fall into two categories: linear or ordinary least squares and non-linear least squares, depending on whether or not the residuals are linear in all unknowns. The linear least-squares problem occurs in statistical regression analysis; it has a closed-form solution. The non-linear problem is usually solved by iterative refinement; at each iteration the system is approximated by a linear one, and thus the core calculation is similar in both cases.

Polynomial least squares describes the variance in a prediction of the dependent variable as a function of the independent variable and the deviations from the fitted curve.

When the observations come from an exponential family and mild conditions are satisfied, least-squares estimates and maximum-likelihood estimates are identical. The method of least squares can also be derived as a method of moments estimator.

The following discussion is mostly presented in terms of linear functions but the use of least-squares is valid and practical for more general families of functions. Also, by iteratively applying local quadratic approximation to the likelihood (through the Fisher information), the least-squares method may be used to fit a generalized linear model.

For the topic of approximating a function by a sum of others using an objective function based on squared distances.

The least-squares method is usually credited to Carl Friedrich Gauss (1795), but it was first published by Adrien-Marie Legendre.

History

Context

The method of least squares grew out of the fields of astronomy and geodesy as scientists and mathematicians sought to provide solutions to the challenges of navigating the Earth's oceans during the Age of Exploration. The accurate description of the behavior of celestial bodies was the key to enabling ships to sail in open seas, where sailors could no longer rely on land sightings for navigation.

The method was the culmination of several advances that took place during the course of the eighteenth century:

- The combination of different observations as being the best estimate of the true value; errors decrease with aggregation rather than increase, perhaps first expressed by Roger Cotes in 1722.

- The combination of different observations taken under the *same* conditions contrary to simply trying one's best to observe and record a single observation accurately. The approach was known as the method of averages. This approach was notably used by Tobias Mayer while studying the librations of the moon in 1750, and by Pierre-Simon Laplace in his work in explaining the differences in motion of Jupiter and Saturn in 1788.

- The combination of different observations taken under *different* conditions. The method came to be known as the method of least absolute deviation. It was notably performed by Roger Joseph Boscovich in his work on the shape of the earth in 1757 and by Pierre-Simon Laplace for the same problem in 1799.

- The development of a criterion that can be evaluated to determine when the solution with the minimum error has been achieved. Laplace tried to specify a mathematical form of the probability density for the errors and define a method of estimation that minimizes the error of estimation. For this purpose, Laplace used a symmetric two-sided exponential distribution we now call Laplace distribution to model the error distribution, and used the sum of absolute deviation as error of estimation. He felt these to be the simplest assumptions he could make, and he had hoped to obtain the arithmetic mean as the best estimate. Instead, his estimator was the posterior median.

The Method

Carl Friedrich Gauss

The first clear and concise exposition of the method of least squares was published by Legendre in 1805. The technique is described as an algebraic procedure for fitting linear equations to data and Legendre demonstrates the new method by analyzing the same data as Laplace for the shape of the earth. The value of Legendre's method of least squares was immediately recognized by leading astronomers and geodesists of the time.

In 1809 Carl Friedrich Gauss published his method of calculating the orbits of celestial bodies. In that work he claimed to have been in possession of the method of least squares since 1795. This naturally led to a priority dispute with Legendre. However, to Gauss's credit, he went beyond Legendre and succeeded in connecting the method of least squares with the principles of probability and to the normal distribution. He had managed to complete Laplace's program of specifying a mathematical form of the probability density for the observations, depending on a finite number of unknown parameters, and define a method of estimation that minimizes the error of estimation. Gauss showed that arithmetic mean is indeed the best estimate of the location parameter by changing both the probability density and the method of estimation. He then turned the problem around by asking what form the density should have and what method of estimation should be used to get the arithmetic mean as estimate of the location parameter. In this attempt, he invented the normal distribution.

An early demonstration of the strength of Gauss' Method came when it was used to predict the future location of the newly discovered asteroid Ceres. On 1 January 1801, the Italian astronomer Giuseppe Piazzi discovered Ceres and was able to track its path for 40 days before it was lost in the glare of the sun. Based on these data, astronomers desired to determine the location of Ceres after it emerged from behind the sun without solving Kepler's complicated nonlinear equations of planetary motion. The only predictions that successfully allowed Hungarian astronomer Franz Xaver von Zach to relocate Ceres were those performed by the 24-year-old Gauss using least-squares analysis.

In 1810, after reading Gauss's work, Laplace, after proving the central limit theorem, used it to give a large sample justification for the method of least square and the normal distribution. In 1822, Gauss was able to state that the least-squares approach to regression analysis is optimal in the sense that in a linear model where the errors have a mean of zero, are uncorrelated, and have equal variances, the best linear unbiased estimator of the coefficients is the least-squares estimator. This result is known as the Gauss–Markov theorem.

The idea of least-squares analysis was also independently formulated by the American Robert Adrain in 1808. In the next two centuries workers in the theory of errors and in statistics found many different ways of implementing least squares.

Problem Statement

The objective consists of adjusting the parameters of a model function to best fit a data set. A simple data set consists of n points (data pairs) (x_i, y_i), $i = 1, ..., n$, where $|x_i$ is an independent variable and y_i is a dependent variable whose value is found by observation. The model function has the form $f(x, \beta)$, where m adjustable parameters are held in the vector β. The goal is to find the parameter values for the model that "best" fits the data. The least squares method finds its optimum when the sum, S, of squared residuals

$$S = \sum_{i=1}^{n} r_i^2$$

is a minimum. A residual is defined as the difference between the actual value of the dependent variable and the value predicted by the model. Each data point has one residual. Both the sum and the mean of the residuals are equal to zero.

$$r_i = y_i - f(x_i, \beta).$$

An example of a model is that of the straight line in two dimensions. Denoting the y-intercept as β_0 and the slope as β_1, the model function is given by $f(x, \beta) = \beta_0 + \beta_1 x$.

A data point may consist of more than one independent variable. For example, when fitting a plane to a set of height measurements, the plane is a function of two independent variables, x and z, say. In the most general case there may be one or more independent variables and one or more dependent variables at each data point.

Limitations

This regression formulation considers only residuals in the dependent variable. There are two rather different contexts in which different implications apply:

- Regression for prediction. Here a model is fitted to provide a prediction rule for application in a similar situation to which the data used for fitting apply. Here the dependent variables corresponding to such future application would be subject to the same types of observation error as those in the data used for fitting. It is therefore logically consistent to use the least-squares prediction rule for such data.

- Regression for fitting a "true relationship". In standard regression analysis, that leads to fitting by least squares, there is an implicit assumption that errors in the independent variable are zero or strictly controlled so as to be negligible. When errors in the independent variable are non-negligible, models of measurement error can be used; such methods can lead to parameter estimates, hypothesis testing and confidence intervals that take into account the presence of observation errors in the independent variables. An alternative approach is to fit a model by total least squares; this can be viewed as taking a pragmatic approach to balancing the effects of the different sources of error in formulating an objective function for use in model-fitting.

Solving the Least Squares Problem

The minimum of the sum of squares is found by setting the gradient to zero. Since the model contains m parameters, there are m gradient equations:

$$\frac{\partial S}{\partial \beta_j} = 2\sum_i r_i \frac{\partial r_i}{\partial \beta_j} = 0, \; j = 1, \ldots, m$$

and since $r_i = y_i - f(x_i, \beta)$,, the gradient equations become

$$-2\sum_i r_i \frac{\partial f(x_i, \beta)}{\partial \beta_j} = 0, \; j = 1, \ldots, m.$$

The gradient equations apply to all least squares problems. Each particular problem requires particular expressions for the model and its partial derivatives.

Linear Least Squares

A regression model is a linear one when the model comprises a linear combination of the parameters, i.e.,

$$f(x, \beta) = \sum_{j=1}^{m} \beta_j \phi_j(x),$$

where the function ϕ_j is a function of x.

Letting

$$X_{ij} = \frac{\partial f(x_i, \beta)}{\partial \beta_j} = \phi_j(x_i),$$

we can then see that in that case the least square estimate (or estimator, in the context of a random sample), β is given by

$$\hat{\beta} = (X^T X)^{-1} X^T y.$$

For a derivation of this estimate see Linear least squares (mathematics).

Non-linear Least Squares

There is no closed-form solution to a non-linear least squares problem. Instead, numerical algorithms are used to find the value of the parameters β that minimizes the objective. Most algorithms involve choosing initial values for the parameters. Then, the parameters are refined iteratively, that is, the values are obtained by successive approximation:

$$\beta_j^{k+1} = \beta_j^k + \Delta\beta_j,$$

where k is an iteration number, and the vector of increments $\Delta\beta_j$ is called the shift vector. In some commonly used algorithms, at each iteration the model may be linearized by approximation to a first-order Taylor series expansion about β^k:

$$\begin{aligned} f(x_i, \beta) &= f^k(x_i, \beta) + \sum_j \frac{\partial f(x_i, \beta)}{\partial \beta_j}\left(\beta_j - \beta_j^k\right) \\ &= f^k(x_i, \beta) + \sum_j J_{ij}\Delta\beta_j. \end{aligned}$$

The Jacobian J is a function of constants, the independent variable *and* the parameters, so it changes from one iteration to the next. The residuals are given by

$$r_i = y_i - f^k(x_i, \beta) - \sum_{k=1}^m J_{ik}\Delta\beta_k = \Delta y_i - \sum_{j=1}^m J_{ij}\Delta\beta_j.$$

To minimize the sum of squares of r_i, the gradient equation is set to zero and solved for $\Delta\beta_j$:

$$-2\sum_{i=1}^n J_{ij}\left(\Delta y_i - \sum_{k=1}^m J_{ik}\Delta\beta_k\right) = 0,$$

which, on rearrangement, become m simultaneous linear equations, the normal equations:

$$\sum_{i=1}^n \sum_{k=1}^m J_{ij}J_{ik}\Delta\beta_k = \sum_{i=1}^n J_{ij}\Delta y_i \qquad (j = 1, \ldots, m)$$

The normal equations are written in matrix notation as

$$\left(J^T J\right)\Delta\beta = J^T \Delta Y.$$

These are the defining equations of the Gauss–Newton algorithm.

Differences between Linear and Nonlinear Least Squares

- The model function, f, in LLSQ (linear least squares) is a linear combination of parameters of the form $f = X_{i1}\beta_1 + X_{i2}\beta_2 + \cdots$ The model may represent a straight line, a parabola or any other linear combination of functions. In NLLSQ (nonlinear least squares) the parameters appear as functions, such as $\beta^2, e^{\beta x}$ and so forth. If the derivatives $\partial f / \partial \beta_j$ are either constant or depend only on the values of the independent variable, the model is linear in the parameters. Otherwise the model is nonlinear.

- Algorithms for finding the solution to a NLLSQ problem require initial values for the parameters, LLSQ does not.

- Like LLSQ, solution algorithms for NLLSQ often require that the Jacobian be calculated. Analytical expressions for the partial derivatives can be complicated. If analytical expressions are impossible to obtain either the partial derivatives must be calculated by numerical approximation or an estimate must be made of the Jacobian.

- In NLLSQ non-convergence (failure of the algorithm to find a minimum) is a common phenomenon whereas the LLSQ is globally concave so non-convergence is not an issue.

- NLLSQ is usually an iterative process. The iterative process has to be terminated when a convergence criterion is satisfied. LLSQ solutions can be computed using direct methods, although problems with large numbers of parameters are typically solved with iterative methods, such as the Gauss–Seidel method.

- In LLSQ the solution is unique, but in NLLSQ there may be multiple minima in the sum of squares.

- Under the condition that the errors are uncorrelated with the predictor variables, LLSQ yields unbiased estimates, but even under that condition NLLSQ estimates are generally biased.

These differences must be considered whenever the solution to a nonlinear least squares problem is being sought.

Least Squares, Regression Analysis and Statistics

The method of least squares is often used to generate estimators and other statistics in regression analysis.

Consider a simple example drawn from physics. A spring should obey Hooke's law which states that the extension of a spring y is proportional to the force, F, applied to it.

$$y = f(F, k) = kF$$

constitutes the model, where F is the independent variable. To estimate the force constant, k, a series of n measurements with different forces will produce a set of data,

$(F_i, y_i), i = 1, \ldots, n$, where y_i is a measured spring extension. Each experimental observation will contain some error. If we denote this error ε, we may specify an empirical model for our observations,

$$y_i = kF_i + \varepsilon_i.$$

There are many methods we might use to estimate the unknown parameter k. Noting that the n equations in the m variables in our data comprise an overdetermined system with one unknown and n equations, we may choose to estimate k using least squares. The sum of squares to be minimized is

$$S = \sum_{i=1}^{n} (y_i - kF_i)^2$$

The least squares estimate of the force constant, k, is given by

$$\hat{k} = \frac{\sum_i F_i y_i}{\sum_i F_i^2}.$$

Here it is assumed that application of the force *causes* the spring to expand and, having derived the force constant by least squares fitting, the extension can be predicted from Hooke's law.

In regression analysis the researcher specifies an empirical model. For example, a very common model is the straight line model which is used to test if there is a linear relationship between dependent and independent variable. If a linear relationship is found to exist, the variables are said to be correlated. However, correlation does not prove causation, as both variables may be correlated with other, hidden, variables, or the dependent variable may "reverse" cause the independent variables, or the variables may be otherwise spuriously correlated. For example, suppose there is a correlation between deaths by drowning and the volume of ice cream sales at a particular beach. Yet, both the number of people going swimming and the volume of ice cream sales increase as the weather gets hotter, and presumably the number of deaths by drowning is correlated with the number of people going swimming. Perhaps an increase in swimmers causes both the other variables to increase.

In order to make statistical tests on the results it is necessary to make assumptions about the nature of the experimental errors. A common (but not necessary) assumption is that the errors belong to a normal distribution. The central limit theorem supports the idea that this is a good approximation in many cases.

- The Gauss–Markov theorem. In a linear model in which the errors have expectation zero conditional on the independent variables, are uncorrelated and have equal variances, the best linear unbiased estimator of any linear combination of the observations, is its least-squares estimator. "Best" means that the least squares estimators of the parameters have minimum variance. The assumption of equal variance is valid when the errors all belong to the same distribution.

- In a linear model, if the errors belong to a normal distribution the least squares estimators are also the maximum likelihood estimators.

However, if the errors are not normally distributed, a central limit theorem often nonetheless implies that the parameter estimates will be approximately normally distributed so long as the sample is reasonably large. For this reason, given the important property that the error mean is independent of the independent variables, the distribution of the error term is not an important issue in regression analysis. Specifically, it is not typically important whether the error term follows a normal distribution.

In a least squares calculation with unit weights, or in linear regression, the variance on the jth parameter, denoted $\mathrm{var}(\hat{\beta}_j)$, , is usually estimated with

$$\mathrm{var}(\hat{\beta}_j) = \sigma^2 \left(\left[X^T X \right]^{-1} \right)_{jj} \approx \frac{S}{n-m} \left(\left[X^T X \right]^{-1} \right)_{jj},$$

where the true residual variance σ^2 is replaced by an estimate based on the minimised value of the sum of squares objective function S. The denominator, $n - m$, is the statistical degrees of freedom.

Confidence limits can be found if the probability distribution of the parameters is known, or an asymptotic approximation is made, or assumed. Likewise statistical tests on the residuals can be made if the probability distribution of the residuals is known or assumed. The probability distribution of any linear combination of the dependent variables can be derived if the probability distribution of experimental errors is known or assumed. Inference is particularly straightforward if the errors are assumed to follow a normal distribution, which implies that the parameter estimates and residuals will also be normally distributed conditional on the values of the independent variables.

Weighted Least Squares

A special case of generalized least squares called weighted least squares occurs when all the off-diagonal entries of Ω (the correlation matrix of the residuals) are null; the variances of the observations (along the covariance matrix diagonal) may still be unequal (heteroscedasticity).

The expressions given above are based on the implicit assumption that the errors are uncorrelated with each other and with the independent variables and have equal variance. The Gauss–Markov theorem shows that, when this is so, $\hat{\beta}$ is a best linear unbiased estimator (BLUE). If, however, the measurements are uncorrelated but have different uncertainties, a modified approach might be adopted. Aitken showed that when a weighted sum of squared residuals is minimized, $\hat{\beta}$ is the BLUE if each weight is equal to the reciprocal of the variance of the measurement

$$S = \sum_{i=1}^{n} W_{ii} r_i^2, \qquad W_{ii} = \frac{1}{\sigma_i^2}$$

The gradient equations for this sum of squares are

$$-2 \sum_{i} W_{ii} \frac{\partial f(x_i, \beta)}{\partial \beta_j} r_i = 0, \qquad j = 1, \ldots, n$$

which, in a linear least squares system give the modified normal equations,

$$\sum_{i=1}^{n} \sum_{k=1}^{m} X_{ij} W_{ii} X_{ik} \hat{\beta}_k = \sum_{i=1}^{n} X_{ij} W_{ii} y_i, \qquad j = 1, \ldots, m.$$

When the observational errors are uncorrelated and the weight matrix, W, is diagonal, these may be written as

$$\left(X^T W X \right) \hat{\beta} = X^T W y.$$

If the errors are correlated, the resulting estimator is the BLUE if the weight matrix is equal to the inverse of the variance-covariance matrix of the observations.

When the errors are uncorrelated, it is convenient to simplify the calculations to factor the weight matrix as $w_{ii} = \sqrt{W_{ii}}$. The normal equations can then be written in the same form as ordinary least squares:

$$\left(X'^T X' \right) \hat{\beta} = X'^T y'$$

where we define the following scaled matrix and vector:

$$X' = \operatorname{diag}(w) X,$$
$$y' = \operatorname{diag}(w) y = y \oslash \sigma.$$

For non-linear least squares systems a similar argument shows that the normal equations should be modified as follows.

$$\left(J^T W J \right) \Delta \beta = J^T W \Delta y.$$

Note that for empirical tests, the appropriate W is not known for sure and must be estimated. For this feasible generalized least squares (FGLS) techniques may be used.

Relationship to Principal Components

The first principal component about the mean of a set of points can be represented by that line which most closely approaches the data points (as measured by squared distance of closest approach, i.e. perpendicular to the line). In contrast, linear least squares tries to minimize the distance in the y direction only. Thus, although the two use a similar error metric, linear least squares is a method that treats one dimension of the data preferentially, while PCA treats all dimensions equally.

Regularized Versions

Tikhonov Regularization

In some contexts a regularized version of the least squares solution may be preferable. Tikhonov regularization (or ridge regression) adds a constraint that $\| \beta \|^2$, , the L_2-norm of the parameter vector, is not greater than a given value. Equivalently, it may solve an unconstrained minimization of the least-squares penalty with $\alpha \| \beta \|^2$ added, where α is a constant (this is the Lagrangian form of the constrained problem). In a Bayesian context, this is equivalent to placing a zero-mean normally distributed prior on the parameter vector.

Lasso Method

An alternative regularized version of least squares is *Lasso* (least absolute shrinkage and selection operator), which uses the constraint that $\| \beta \|$, the L_1-norm of the parameter vector, is no greater than a given value. (As above, this is equivalent to an unconstrained minimization of the least-squares penalty with $\alpha \| \beta \|^2$ added.) In a Bayesian context, this is equivalent to placing a zero-mean Laplace prior distribution on the parameter vector. The optimization problem may be solved using quadratic programming or more general convex optimization methods, as well as by specific algorithms such as the least angle regression algorithm.

One of the prime differences between Lasso and ridge regression is that in ridge regression, as the penalty is increased, all parameters are reduced while still remaining non-zero, while in Lasso, increasing the penalty will cause more and more of the parameters to be driven to zero. This is an advantage of Lasso over ridge regression, as driving parameters to zero deselects the features from the regression. Thus, Lasso automatically selects more relevant features and discards the others, whereas Ridge regression never fully discards any features. Some feature selection techniques are developed based on the LASSO including Bolasso which bootstraps samples, and FeaLect which analyzes the regression coefficients corresponding to different values of α to score all the features.

The L^1-regularized formulation is useful in some contexts due to its tendency to prefer solutions with fewer nonzero parameter values, effectively reducing the number of variables upon which the given solution is dependent. For this reason, the Lasso and its variants are fundamental to the field of compressed sensing. An extension of this approach is elastic net regularization.

Probability Distribution

In probability and statistics, a probability distribution is a mathematical description of a random phenomenon in terms of the probabilities of events. Examples of random phenomena include the results of an experiment or survey. A probability distribution is defined in terms of an underlying sample space, which is the set of all possible outcomes of the random phenomenon being observed. The sample space may be the set of real numbers or a higher-dimensional vector space, or it may be a list of non-numerical values; for example, the sample space of a coin flip would be . Probability distributions are generally divided into two classes. A discrete probability distribution can be encoded by a list of the probabilities of the outcomes, known as a probability mass function. On the other hand, in a continuous probability distribution, the probability of any individual outcome is 0. Continuous probability distributions can often be described by probability density functions; however, more complex experiments, such as those involving stochastic processes defined in continuous time, may demand the use of more general probability measures.

In applied probability, a probability distribution can be specified in a number of different ways, often chosen for mathematical convenience:

- by supplying a valid probability mass function or probability density function

- by supplying a valid cumulative distribution function or survival function

- by supplying a valid hazard function

- by supplying a valid characteristic function

- by supplying a rule for constructing a new random variable from other random variables whose joint probability distribution is known.

A probability distribution whose sample space is the set of real numbers is called univariate, while a distribution whose sample space is a vector space is called multivariate. A univariate distribution gives the probabilities of a single random variable taking on various alternative values; a multivariate distribution (a joint probability distribution) gives the probabilities of a random vector—a list of two or more random variables—taking on various combinations of values. Important and commonly encountered univariate probability distributions include the binomial distribution, the hypergeometric distribution, and the normal distribution. The multivariate normal distribution is a commonly encountered multivariate distribution.

Introduction

To define probability distributions for the simplest cases, one needs to distinguish between discrete and continuous random variables. In the discrete case, it is sufficient to specify a probability mass function assigning a probability to each possible outcome: for example, when throwing a fair dice, each of the six values *1* to *6* has the probability 1/6. The probability of an event is then defined to be the sum of the probabilities of the outcomes that satisfy the event; for example, the probability of the event "the die rolls an even value" is

$$Prob(2) + Prob(4) + Prob(6) = 1/6 + 1/6 + 1/6 = 1/2.$$

In contrast, when a random variable takes values from a continuum then typically, any individual outcome has probability zero and only events that include infinitely many outcomes, such as intervals, can have positive probability. For example, the probability that a given object weighs *exactly* 500 g is zero, because the probability of measuring exactly 500 g tends to zero as the accuracy of our measuring instruments increases. Nevertheless, in quality control one might demand that the probability of a "500 g" package containing between 490 g and 510 g should be no less than 98%, and this demand is less sensitive to the accuracy of our instruments.

Continuous probability distributions can be described in several ways. The probability density function describes the infinitesimal probability of any given value, and the probability that the outcome lies in a given interval can be computed by integrating the probability density function over that interval. On the other hand, the cumulative distribution function describes the probability that the random variable is no larger than a given value; the probability that the outcome lies in a given interval can be computed by taking the difference between the values of the cumulative distribution function at the endpoints of the interval. The cumulative distribution function is the antiderivative of the probability density function provided that the latter function exists.

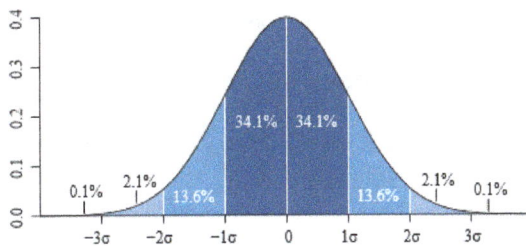

The probability density function (pdf) of the normal distribution, also called Gaussian or "bell curve", the most important continuous random distribution. As noted on the figure, the probabilities of intervals of values correspond to the area under the curve.

Terminology

As probability theory is used in quite diverse applications, terminology is not uniform and sometimes confusing. The following terms are used for non-cumulative probability distribution functions:

- Probability mass, Probability mass function, p.m.f.: for discrete random variables.

- Categorical distribution: for discrete random variables with a finite set of values.

- Probability density, Probability density function, p.d.f.: most often reserved for continuous random variables.

The following terms are somewhat ambiguous as they can refer to non-cumulative or cumulative distributions, depending on authors' preferences:

- Probability distribution function: continuous or discrete, non-cumulative or cumulative.

- Probability function: even more ambiguous, can mean any of the above or other things.

Finally,

- Probability distribution: sometimes the same as *probability distribution function*, but usually refers to the more complete assignment of probabilities to all measurable subsets of outcomes (i.e. the corresponding probability measure), not just to specific outcomes or ranges of outcomes.

Basic Terms

- Mode: for a discrete random variable, the value with highest probability (the location at which the probability mass function has its peak); for a continuous random variable, the location at which the probability density function has its peak.

- Support: the smallest closed set whose complement has probability zero.

- Head: the range of values where the pmf or pdf is relatively high.

- Tail: the complement of the head within the support; the large set of values where the pmf or pdf is relatively low.

- Expected value or mean: the weighted average of the possible values, using their probabilities as their weights; or the continuous analog thereof.

- Median: the value such that the set of values less than the median has a probability of one-half.

- Variance: the second moment of the pmf or pdf about the mean; an important measure of the dispersion of the distribution.

- Standard deviation: the square root of the variance, and hence another measure of dispersion.

- Symmetry: a property of some distributions in which the portion of the distribution to the left of a specific value is a mirror image of the portion to its right.

- Skewness: a measure of the extent to which a pmf or pdf "leans" to one side of its mean. The third standardized moment of the distribution.

- Kurtosis: a measure of the "fatness" of the tails of a pmf or pdf. The fourth standardized moment of the distribution.

Cumulative Distribution Function

Because a probability distribution Pr on the real line is determined by the probability of a scalar random variable X being in a half-open interval $(-\infty, x]$, the probability distribution is completely characterized by its cumulative distribution function:

$$F(x) = Pr\left[X \leq x\right] \qquad \text{for all } x \in \mathbb{R}.$$

Discrete Probability Distribution

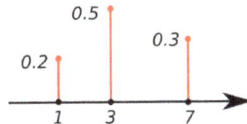

The probability mass function of a discrete probability distribution. The probabilities of the singletons {1}, {3}, and {7} are respectively 0.2, 0.5, 0.3. A set not containing any of these points has probability zero.

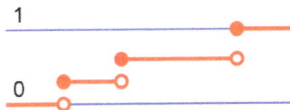

The cdf of a discrete probability distribution, ...

... of a continuous probability distribution, ...

... of a distribution which has both a continuous part and a discrete part.

A discrete probability distribution is a probability distribution characterized by a probability mass function. Thus, the distribution of a random variable X is discrete, and X is called a discrete random variable, if

$$\sum_{u} \Pr(X = u) = 1$$

as u runs through the set of all possible values of X. A discrete random variable can assume only a finite or countably infinite number of values. For the number of potential values to be countably infinite, even though their probabilities sum to 1, the probabilities have to decline to zero fast enough. For example, if $\Pr(X = n) = \frac{1}{2^n}$ for $n = 1, 2, ...,$ we have the sum of probabilities $1/2 + 1/4 + 1/8 + ... = 1$.

Well-known discrete probability distributions used in statistical modeling include the Poisson distribution, the Bernoulli distribution, the binomial distribution, the geometric distribution, and the negative binomial distribution. Additionally, the discrete uniform distribution is commonly used in computer programs that make equal-probability random selections between a number of choices.

Measure Theoretic Formulation

A measurable function $X : A \to B$ between a probability space $(A, \mathcal{A}, P)$ and a measurable space $(B, \mathcal{B})$ is called a discrete random variable provided its image is a countable set and the pre-image of singleton sets are measurable, i.e., $X^{-1}(b) \in \mathcal{A}$ for all $b \in B.$. The latter requirement induces a probability mass function $f_x : X(A) \to \mathbb{R}$ via $f_x(b) := P(X^{-1}(b))$. Since the pre-images of disjoint sets are disjoint

$$\sum_{b \in X(A)} f_x(b) = \sum_{b \in X(A)} P(X^{-1}(b)) = P\left(\bigcup_{b \in X(A)} X^{-1}(b) \right) = P(A) = 1.$$

This recovers the definition given above.

Cumulative Density

Equivalently to the above, a discrete random variable can be defined as a random variable whose cumulative distribution function (cdf) increases only by jump discontinuities—that is, its cdf increases only where it "jumps" to a higher value, and is constant between those jumps. The points where jumps occur are precisely the values which the random variable may take.

Delta-function Representation

Consequently, a discrete probability distribution is often represented as a generalized probability density function involving Dirac delta functions, which substantially unifies the treatment of continuous and discrete distributions. This is especially useful when dealing with probability distributions involving both a continuous and a discrete part.

Indicator-function Representation

For a discrete random variable X, let $u_0, u_1, ...$ be the values it can take with non-zero probability. Denote

$$\Omega_i = X^{-1}(u_i) = \{\omega : X(\omega) = u_i\}, i = 0,1,2,\ldots$$

These are disjoint sets, and by formula (1)

$$\Pr\left(\bigcup_i \Omega_i\right) = \sum_i \Pr(\Omega_i) = \sum_i \Pr(X = u_i) = 1.$$

It follows that the probability that X takes any value except for $u_0, u_1, \ldots$ is zero, and thus one can write X as

$$X(\omega) = \sum_i u_i 1_{\Omega_i}(\omega)$$

except on a set of probability zero, where 1_A is the indicator function of A. This may serve as an alternative definition of discrete random variables.

Continuous Probability Distribution

A continuous probability distribution is a *probability distribution* that has a cumulative distribution function that is continuous. Most often they are generated by having a probability density function. Mathematicians call distributions with probability density functions absolutely continuous, since their cumulative distribution function is absolutely continuous with respect to the Lebesgue measure λ. If the distribution of X is continuous, then X is called a continuous random variable. There are many examples of continuous probability distributions: normal, uniform, chi-squared, and others.

Intuitively, a continuous random variable is the one which can take a continuous range of values—as opposed to a discrete distribution, where the set of possible values for the random variable is at most countable. While for a discrete distribution an event with probability zero is impossible (e.g., rolling 31/2 on a standard dice is impossible, and has probability zero), this is not so in the case of a continuous random variable. For example, if one measures the width of an oak leaf, the result of 3½ cm is possible; however, it has probability zero because uncountably many other potential values exist even between 3 cm and 4 cm. Each of these individual outcomes has probability zero, yet the probability that the outcome will fall into the interval (3 cm, 4 cm) is nonzero. This apparent paradox is resolved by the fact that the probability that X attains some value within an infinite set, such as an interval, cannot be found by naively adding the probabilities for individual values. Formally, each value has an infinitesimally small probability, which statistically is equivalent to zero.

Formally, if X is a continuous random variable, then it has a probability density function $f(x)$, and therefore its probability of falling into a given interval, say $[a, b]$ is given by the integral

$$\Pr[a \le X \le b] = \int_a^b f(x)dx$$

In particular, the probability for X to take any single value a (that is $a \le X \le a$) is zero, because an integral with coinciding upper and lower limits is always equal to zero.

The definition states that a continuous probability distribution must possess a density, or equivalently, its cumulative distribution function be absolutely continuous. This requirement is stronger than simple continuity of the cumulative distribution function, and there is a special class of distributions, singular distributions, which are neither continuous nor discrete nor a mixture of those. An example is given by the Cantor distribution. Such singular distributions however are never encountered in practice.

Note on terminology: some authors use the term "continuous distribution" to denote the distribution with continuous cumulative distribution function. Thus, their definition includes both the (absolutely) continuous and singular distributions.

By one convention, a probability distribution μ is called *continuous* if its cumulative distribution function $F(x) = \mu(-\infty, x]$ is continuous and, therefore, the probability measure of singletons $\mu\{x\}=0$ for all x.

Another convention reserves the term *continuous probability distribution* for absolutely continuous distributions. These distributions can be characterized by a probability density function: a non-negative Lebesgue integrable function f defined on the real numbers such that

$$F(x) = \mu(-\infty, x] = \int_{-\infty}^x f(t)dt.$$

Discrete distributions and some continuous distributions (like the Cantor distribution) do not admit such a density.

Some Properties

- The probability distribution of the sum of two independent random variables is the convolution of each of their distributions.

- Probability distributions are not a vector space—they are not closed under linear combinations, as these do not preserve non-negativity or total integral 1— but they are closed under convex combination, thus forming a convex subset of the space of functions (or measures).

Kolmogorov Definition

In the measure-theoretic formalization of probability theory, a random variable is defined as a measurable function X from a probability space $(\Omega, \mathcal{F}, P)$ to measurable space

$(\mathcal{X}, \mathcal{A})$.. A probability distribution of X is the pushforward measure X_*P of X, which is a probability measure on $(\mathcal{X}, \mathcal{A})$ satisfying $X_*P = PX^{-1}$.

Random Number Generation

A frequent problem in statistical simulations (the Monte Carlo method) is the generation of pseudo-random numbers that are distributed in a given way. Most algorithms are based on a pseudorandom number generator that produces numbers X that are uniformly distributed in the interval [0,1). These random variates X are then transformed via some algorithm to create a new random variate having the required probability distribution.

Applications

The concept of the probability distribution and the random variables which they describe underlies the mathematical discipline of probability theory, and the science of statistics. There is spread or variability in almost any value that can be measured in a population (e.g. height of people, durability of a metal, sales growth, traffic flow, etc.); almost all measurements are made with some intrinsic error; in physics many processes are described probabilistically,from the kinetic properties of gases to the quantum mechanical description of fundamental particles. For these and many other reasons, simple numbers are often inadequate for describing a quantity, while probability distributions are often more appropriate.

As a more specific example of an application, the cache language models and other statistical language models used in natural language processing to assign probabilities to the occurrence of particular words and word sequences do so by means of probability distributions.

Common Probability Distributions

The following is a list of some of the most common probability distributions, grouped by the type of process that they are related to.

Note also that all of the univariate distributions below are singly peaked; that is, it is assumed that the values cluster around a single point. In practice, actually observed quantities may cluster around multiple values. Such quantities can be modeled using a mixture distribution.

Related to Real-valued Quantities that Grow Linearly (e.g. Errors, Offsets)

- Normal distribution (Gaussian distribution), for a single such quantity; the most common continuous distribution

Related to Positive Real-valued Quantities that Grow Exponentially (e.g. Prices, Incomes, Populations)

- Log-normal distribution, for a single such quantity whose log is normally distributed

- Pareto distribution, for a single such quantity whose log is exponentially distributed; the prototypical power law distribution

Related to Real-valued Quantities that are Assumed to be Uniformly Distributed Over a (Possibly Unknown) Region

- Discrete uniform distribution, for a finite set of values (e.g. the outcome of a fair die)

- Continuous uniform distribution, for continuously distributed values

Related to Bernoulli Trials (yes/no Events, with a Given Probability)

- Basic distributions:

  o Bernoulli distribution, for the outcome of a single Bernoulli trial (e.g. success/failure, yes/no)

  o Binomial distribution, for the number of "positive occurrences" (e.g. successes, yes votes, etc.) given a fixed total number of independent occurrences

  o Negative binomial distribution, for binomial-type observations but where the quantity of interest is the number of failures before a given number of successes occurs

  o Geometric distribution, for binomial-type observations but where the quantity of interest is the number of failures before the first success; a special case of the negative binomial distribution

- Related to sampling schemes over a finite population:

  o Hypergeometric distribution, for the number of "positive occurrences" (e.g. successes, yes votes, etc.) given a fixed number of total occurrences, using sampling without replacement

  o Beta-binomial distribution, for the number of "positive occurrences" (e.g. successes, yes votes, etc.) given a fixed number of total occurrences, sampling using a Polya urn scheme (in some sense, the "opposite" of sampling without replacement)

Related to Categorical Outcomes (Events with *K* Possible Outcomes, with a Given Probability for Each Outcome)

- Categorical distribution, for a single categorical outcome (e.g. yes/no/maybe in a survey); a generalization of the Bernoulli distribution

- Multinomial distribution, for the number of each type of categorical outcome, given a fixed number of total outcomes; a generalization of the binomial distribution

- Multivariate hypergeometric distribution, similar to the multinomial distribution, but using sampling without replacement; a generalization of the hypergeometric distribution

Related to Events in a Poisson process (Events that Occur Independently with a given rate)

- Poisson distribution, for the number of occurrences of a Poisson-type event in a given period of time

- Exponential distribution, for the time before the next Poisson-type event occurs

- Gamma distribution, for the time before the next k Poisson-type events occur

Related to the Absolute Values of Vectors with Normally Distributed Components

- Rayleigh distribution, for the distribution of vector magnitudes with Gaussian distributed orthogonal components. Rayleigh distributions are found in RF signals with Gaussian real and imaginary components.

- Rice distribution, a generalization of the Rayleigh distributions for where there is a stationary background signal component. Found in Rician fading of radio signals due to multipath propagation and in MR images with noise corruption on non-zero NMR signals.

Related to Normally Distributed Quantities Operated with sum of Squares (for Hypothesis Testing)

- Chi-squared distribution, the distribution of a sum of squared standard normal variables; useful e.g. for inference regarding the sample variance of normally distributed samples

- Student's t distribution, the distribution of the ratio of a standard normal variable and the square root of a scaled chi squared variable; useful for inference regarding the mean of normally distributed samples with unknown variance

- F-distribution, the distribution of the ratio of two scaled chi squared variables; useful e.g. for inferences that involve comparing variances or involving R-squared (the squared correlation coefficient)

Useful as Conjugate prior Distributions in Bayesian Inference

- Beta distribution, for a single probability (real number between 0 and 1); conjugate to the Bernoulli distribution and binomial distribution

- Gamma distribution, for a non-negative scaling parameter; conjugate to the rate parameter of a Poisson distribution or exponential distribution, the precision (inverse variance) of a normal distribution, etc.

- Dirichlet distribution, for a vector of probabilities that must sum to 1; conjugate to the categorical distribution and multinomial distribution; generalization of the beta distribution

- Wishart distribution, for a symmetric non-negative definite matrix; conjugate to the inverse of the covariance matrix of a multivariate normal distribution; generalization of the gamma distribution

Foundations of Statistics

Foundations of statistics is the usual name for the epistemological debate in statistics over how one should conduct inductive inference from data. Among the issues considered in statistical inference are the question of Bayesian inference versus frequentist inference, the distinction between Fisher's "significance testing" and Neyman-Pearson "hypothesis testing", and whether the likelihood principle should be followed. Some of these issues have been debated for up to 200 years without resolution.

Bandyopadhyay & Forster describe four statistical paradigms: "(1) classical statistics or error statistics, (ii) Bayesian statistics, (iii) likelihood-based statistics, and (iv) the Akaikean-Information Criterion-based statistics".

Savage's text *Foundations of Statistics* has been cited over 12000 times on Google Scholar. It tells the following.

It is unanimously agreed that statistics depends somehow on probability. But, as to what probability is and how it is connected with statistics, there has seldom been such complete disagreement and breakdown of communication since the Tower of Babel. Doubtless, much of the disagreement is merely terminological and would disappear under sufficiently sharp analysis.

Fisher's "Significance Testing" vs Neyman-Pearson "Hypothesis Testing"

In the development of classical statistics in the second quarter of the 20th century two competing models of inductive statistical testing were developed. Their relative merits were hotly debated (for over 25 years) until Fisher's death. While a hybrid of the two methods is widely taught and used, the philosophical questions raised in the debate have not been resolved.

Significance Testing

Fisher popularized significance testing, primarily in two popular and highly influential books. Fisher's writing style in these books was strong on examples and relatively weak on explanations. The books lacked proofs or derivations of significance test statistics (which placed statistical practice in advance of statistical theory). Fisher's more explanatory and philosophical writing was written much later. There appear to be some differences between his earlier practices and his later opinions.

Fisher was motivated to obtain scientific experimental results without the explicit influence of prior opinion. The significance test is a probabilistic version of Modus tollens, a classic form of deductive inference. The significance test might be simplistically stated, "If the evidence is sufficiently discordant with the hypothesis, reject the hypothesis". In application, a statistic is calculated from the experimental data, a probability of exceeding that statistic is determined and the probability is compared to a threshold. The threshold (the numeric version of "sufficiently discordant") is arbitrary (usually decided by convention). A common application of the method is deciding whether a treatment has a reportable effect based on a comparative experiment. Statistical significance is a measure of probability not practical importance. It can be regarded as a requirement placed on statistical signal/noise. The method is based on the assumed existence of an imaginary infinite population corresponding to the null hypothesis.

The significance test requires only one hypothesis. The result of the test is to reject the hypothesis (or not), a simple dichotomy. The test distinguish between truth of the hypothesis and insufficiency of evidence to to disprove the hypothesis; so it is like a criminal trial in which the defendant's guilt is assessed in (so it is like a criminal trial in which the defendant is assumed innocent until proven guilty).

Hypothesis Testing

Neyman & Pearson collaborated on a different, but related, problem – selecting among competing hypotheses based on the experimental evidence alone. Of their joint papers the most cited was from 1933. The famous result of that paper is the Neyman-Pearson lemma. The lemma says that a ratio of probabilities is an excellent criterion for selecting a hypothesis (with the threshold for comparison being arbitrary). The paper proved an optimality of Student's t-test (one of the significance tests). Neyman expressed the opinion that hypothesis testing was a generalization of and an improvement on significance testing. The rationale for their methods is found in their joint papers.

Hypothesis testing requires multiple hypotheses. A hypothesis is always selected, a multiple choice. A lack of evidence is not an immediate consideration. The method is based on the assumption of a repeated sampling of the same population (the classical frequentist assumption).

Grounds of Disagreement

The length of the dispute allowed the debate of a wide range of issues regarded as foundational to statistics.

An example exchange from 1955-1956		
Fisher's Attack	**Neyman's Rebuttal**	**Discussion**
Repeated sampling of the same population • Such sampling is the basis of frequentist probability • Fisher preferred fiducial inference	Fisher's theory of fiducial inference is flawed • Paradoxes are common	Fisher's attack on the basis of frequentist probability failed, but was not without result. He identified a specific case (2x2 table) where the two schools of testing reach different results. This case is one of several that are still troubling. Commentators believe that the "right" answer is context dependent. Fiducial probability has not fared well, being virtually without advocates, while frequentist probability remains a mainstream interpretation.
Type II errors • Which result from an alternative hypothesis	A purely probabilistic theory of tests requires an alternative hypothesis	Fisher's attack on type II errors has faded with time. In the intervening years statistics has separated the exploratory from the confirmatory. In the current environment, the concept of type II errors is used in power calculations for confirmatory hypothesis test sample size determination.
Inductive behavior • (Vs inductive reasoning, Fisher's preference)		Fisher's attack on inductive behavior has been largely successful because of his selection of the field of battle. While *operational decisions* are routinely made on a variety of criteria (such as cost), *scientific conclusions* from experimentation are typically made on the basis of probability alone.

In this exchange Fisher also discussed the requirements for inductive inference, with specific criticism of cost functions penalizing faulty judgments. Neyman countered that Gauss and Laplace used them. This exchange of arguments occurred 15 years *after* textbooks began teaching a hybrid theory of statistical testing.

Fisher and Neyman were in disagreement about the foundations of statistics (although united in opposition to the Bayesian view):

- The interpretation of probability

    o   The disagreement over Fisher's inductive reasoning vs Neyman's inductive behavior contained elements of the Bayesian/Frequentist divide. Fisher was willing to alter his opinion (reaching a provisional conclusion) on the basis of a calculated probability while Neyman was more willing to change his observable behavior (making a decision) on the basis of a computed cost.

- The proper formulation of scientific questions with special concern for modeling

- Whether it is reasonable to reject a hypothesis based on a low probability without knowing the probability of an alternative

- Whether a hypothesis could ever be accepted on the basis of data

    o   In mathematics, deduction proves, counter-examples disprove

    o   In the Popperian philosophy of science, advancements are made when theories are disproven

- Subjectivity: While Fisher and Neyman struggled to minimize subjectivity, both acknowledged the importance of "good judgment". Each accused the other of subjectivity.

    o   Fisher *subjectively* chose the null hypothesis.

    o   Neyman-Pearson *subjectively* chose the criterion for selection (which was not limited to a probability).

    o   Both *subjectively* determined numeric thresholds.

Fisher and Neyman were separated by attitudes and perhaps language. Fisher was a scientist and an intuitive mathematician. Inductive reasoning was natural. Neyman was a rigorous mathematician. He was convinced by deductive reasoning rather by a probability calculation based on an experiment. Thus there was an underlying clash between applied and theoretical, between science and mathematics.

Related History

Neyman, who had occupied the same building in England as Fisher, accepted a position on the west coast of the United States of America in 1938. His move effectively ended his collaboration with Pearson and their development of hypothesis testing. Further development was continued by others.

Textbooks provided a hybrid version of significance and hypothesis testing by 1940.

None of the principals had any known personal involvement in the further development of the hybrid taught in introductory statistics today.

Statistics later developed in different directions including decision theory (and possibly game theory), Bayesian statistics, exploratory data analysis, robust statistics and nonparametric statistics. Neyman-Pearson hypothesis testing contributed strongly to decision theory which is very heavily used (in statistical quality control for example). Hypothesis testing readily generalized to accept prior probabilities which gave it a Bayesian flavor. Neyman-Pearson hypothesis testing has become an abstract mathematical subject taught in post-graduate statistics, while most of what is taught to under-graduates and used under the banner of hypothesis testing is from Fisher.

Contemporary Opinion

No major battles between the two classical schools of testing have erupted for decades, but sniping continues (perhaps encouraged by partisans of other controversies). After generations of dispute, there is virtually no chance that either statistical testing theory will replace the other in the foreseeable future.

The hybrid of the two competing schools of testing can be viewed very differently – as the imperfect union of two mathematically complementary ideas or as the fundamentally flawed union of philosophically incompatible ideas. Fisher enjoyed some philosophical advantage, while Neyman & Pearson employed the more rigorous mathematics. Hypothesis testing is controversial among some users, but the most popular alternative (confidence intervals) is based on the same mathematics.

The history of the development left testing without a single citable authoritative source for the hybrid theory that reflects common statistical practice. The merged terminology is also somewhat inconsistent. There is strong empirical evidence that the graduates (and instructors) of an introductory statistics class have a weak understanding of the meaning of hypothesis testing.

Summary

- The interpretation of probability has not been resolved (but fiducial probability is an orphan).

- Neither test method has been rejected. Both are heavily used for different purposes.

- Texts have merged the two test methods under the term hypothesis testing.

  o Mathematicians claim (with some exceptions) that significance tests are a special case of hypothesis tests.

  o Others treat the problems and methods as distinct (or incompatible).

- The dispute has adversely affected statistical education.

Bayesian Inference Versus Frequentist Inference

Two different interpretations of probability (based on objective evidence and subjective degrees of belief) have long existed. Gauss and Laplace could have debated alternatives more than 200 years ago. Two competing schools of statistics have developed as a consequence. Classical inferential statistics was largely developed in the second quarter of the 20th Century, much of it in reaction to the (Bayesian) probability of the time which utilized the controversial principle of indifference to establish prior probabilities. The rehabilitation of Bayesian inference was a reaction to the limitations of frequentist probability. More reactions followed. While the philosophical interpretations are old, the statistical terminology is not. The current statistical terms "Bayesian" and "frequentist" stabilized in the second half of the 20th Century. The (philosophical, mathematical, scientific, statistical) terminology is confusing: the "classical" interpretation of probability is Bayesian while "classical" statistics is frequentist. "Frequentist" also has varying interpretations—different in philosophy than in physics.

The nuances of philosophical probability interpretations are discussed elsewhere. In statistics the alternative interpretations *enable* the analysis of different data using different methods based on different models to achieve slightly different goals. Any statistical comparison of the competing schools considers pragmatic criteria beyond the philosophical.

Major Contributors

Two major contributors to frequentist (classical) methods were Fisher and Neyman. Fisher's interpretation of probability was idiosyncratic (but strongly non-Bayesian). Neyman's views were rigorously frequentist. Three major contributors to 20th century Bayesian statistical philosophy, mathematics and methods were de Finetti, Jeffreys and Savage. Savage popularized de Finetti's ideas in the English-speaking world and made Bayesian mathematics rigorous. In 1965, Dennis Lindley's 2-volume work "Introduction to Probability and Statistics from a Bayesian Viewpoint" brought Bayesian methods to a wide audience. Statistics has advanced over the past three generations; The "authoritative" views of the early contributors are not all current.

Contrasting Approaches

Frequentist Inference

Frequentist inference is partially and tersely described above in (Fisher's "significance testing" vs Neyman-Pearson "hypothesis testing"). Frequentist inference combines several different views. The result is capable of supporting scientific conclusions, making operational decisions and estimating parameters with or without confidence intervals. Frequentist inference is based solely on the (one set of) evidence.

Bayesian Inference

A classical frequency distribution describes the probability of the data. The use of Bayes' theorem allows a more abstract concept – the probability of a hypothesis (corresponding to a theory) given the data. The concept was once known as "inverse probability". Bayesian inference updates the probability estimate for a hypothesis as additional evidence is acquired. Bayesian inference is explicitly based on the evidence and prior opinion, which allows it to be based on multiple sets of evidence.

Comparisons of Characteristics

Frequentists and Bayesians use different models of probability. Frequentists often consider parameters to be fixed but unknown while Bayesians assign probability distributions to similar parameters. Consequently, Bayesians speak of probabilities that don't exist for frequentists; A Bayesian speaks of the probability of a theory while a true frequentist can speak only of the consistency of the evidence with the theory. Example: A frequentist does not say that there is a 95% probability that the true value of a parameter lies within a confidence interval, saying instead that 95% of confidence intervals contain the true value.

Efron's comparative adjectives		
	Bayes	**Frequentist**
• Basis	• Belief (prior)	• Behavior (method)
• Resulting Characteristic	• Principled Philosophy	• Opportunistic Methods
• —	• One distribution	• Many distributions (bootstrap?)
• Ideal Appwlication	• Dynamic (repeated sampling)	• Static (one sample)
• Target Audience	• Individual (subjective)	• Community (objective)
• Modeling Characteristic	• Aggressive	• Defensive

Alternative comparison		
	Bayesian	**Frequentist**
	• Complete	• Inferences well calibrated
	• Coherent	• No need to specify prior distributions
Strengths	• Prescriptive	• Flexible range of procedures
	• —	o Unbiasness, sufficiency, ancillarity...
	• —	o Widely applicable and dependable
	• —	o Asymptotic theory
	• —	o Easy to interpret
	• —	o Can be calculated by hand
	• Strong inference from model	• Strong model formulation & assessment

Weaknesses	• Too subjective for scientific inference	• Incomplete
		• Ambiguous
	• Denies the role of randomization for design	• Incoherent
		• Not prescriptive
	• Requires and relies on full specification of a model (likelihood and prior)	• No unified theory
		• (Over?)emphasis on asymptotic properties
	• —	
	• —	• Weak inference from model
	• —	
	• Weak model formulation & assessment	

Mathematical Results

Neither school is immune from mathematical criticism and neither accepts it without a struggle. Stein's paradox (for example) illustrated that finding a "flat" or "uninformative" prior probability distribution in high dimensions is subtle. Bayesians regard that as peripheral to the core of their philosophy while finding frequentism to be riddled with inconsistencies, paradoxes and bad mathematical behavior. Frequentists can explain most. Some of the "bad" examples are extreme situations - such as estimating the weight of a herd of elephants from measuring the weight of one ("Basu's elephants"), which allows no statistical estimate of the variability of weights. The likelihood principle has been a battleground.

Statistical Results

Both schools have achieved impressive results in solving real-world problems. Classical statistics effectively has the longer record because numerous results were obtained with mechanical calculators and printed tables of special statistical functions. Bayesian methods have been highly successful in the analysis of information that is naturally sequentially sampled (radar and sonar). Many Bayesian methods and some recent frequentist methods (such as the bootstrap) require the computational power widely available only in the last several decades.

There is hint that Bayesian philosophy is "book smart" compared to Frequentist "street smarts". Bayesian philosophy has sometimes been silent on shuffling the cards. The "design of experiments" teaches the importance of the source of statistical data. Fisher was a major contributor to the theory.

There is active discussion about combining Bayesian and frequentist methods, but reservations are expressed about the meaning of the results and reducing the diversity of approaches.

Philosophical Results

Bayesians are united in opposition to the limitations of frequentism, but are philosophically divided into numerous camps (empirical, hierarchical, objective, personal, subjective), each with a different emphasis. One (frequentist) philosopher of statistics has noted a retreat from the statistical field to philosophical probability interpretations over the last two generations. There is a perception that successes in Bayesian applications do not justify the supporting philosophy. Bayesian methods often create useful models that are not used for traditional inference and which owe little to philosophy. None of the philosophical interpretations of probability (frequentist or Bayesian) appears robust. The frequentist view is too rigid and limiting while the Bayesian view can be simultaneously objective and subjective, etc.

Illustrative Quotations

- "carefully used, the frequentist approach yields broadly applicable if sometimes clumsy answers"

- "To insist on unbiased [frequentist] techniques may lead to negative (but unbiased) estimates of a variance; the use of p-values in multiple tests may lead to blatant contradictions; conventional 0.95-confidence regions may actually consist of the whole real line. No wonder that mathematicians find it often difficult to believe that conventional statistical methods are a branch of mathematics."

- "Bayesianism is a neat and fully principled philosophy, while frequentism is a grab-bag of opportunistic, individually optimal, methods."

- "in multiparameter problems flat priors can yield very bad answers"

- "[Bayes' rule] says there is a simple, elegant way to combine current information with prior experience in order to state how much is known. It implies that sufficiently good data will bring previously disparate observers to agreement. It makes full use of available information, and it produces decisions having the least possible error rate."

- "Bayesian statistics is about making probability statements, frequentist statistics is about evaluating probability statements."

- "[S]tatisticians are often put in a setting reminiscent of Arrow's paradox, where we are asked to provide estimates that are informative and unbiased and confidence statements that are correct conditional on the data and also on the underlying true parameter." (These are conflicting requirements.)

- "formal inferential aspects are often a relatively small part of statistical analysis"

- "The two philosophies, Bayesian and frequentist, are more orthogonal than antithetical."

Summary

- Bayesian theory has a mathematical advantage

    o Frequentist probability has existence and consistency problems

    o But, finding good priors to apply Bayesian theory remains (very?) difficult

- Both theories have impressive records of successful application

- Neither supporting philosophical interpretation of probability is robust

- There is increasing skepticism of the connection between application and philosophy

- Some statisticians are recommending active collaboration (beyond a cease fire)

The Likelihood Principle

Likelihood is a synonym for probability in common usage. In statistics it is reserved for probabilities that fail to meet the frequentist definition. A probability refers to variable data for a fixed hypothesis while a likelihood refers to variable hypotheses for a fixed set of data. Repeated measurements of a fixed length with a ruler generate a set of observations. Each fixed set of observational conditions is associated with a probability distribution and each set of observations can be interpreted as a sample from that distribution – the frequentist view of probability. Alternatively a set of observations may result from sampling any of a number of distributions (each resulting from a set of observational conditions). The probabilistic relationship between a fixed sample and a variable distribution (resulting from a variable hypothesis) is termed likelihood – a Bayesian view of probability. A set of length measurements may imply readings taken by careful, sober, rested, motivated observers in good lighting.

A likelihood is a probability (or not) by another name which exists because of the limited frequentist definition of probability. Likelihood is a concept introduced and advanced by Fisher for more than 40 years (although prior references to the concept exist and Fisher's support was half-hearted). The concept was accepted and substantially changed by Jeffreys. In 1962 Birnbaum "proved" the likelihood principle from premises acceptable to most statisticians. The "proof" has been disputed by statisticians and philosophers. The principle says that all of the information in a sample is contained in the likelihood function, which is accepted as a valid probability distribution by Bayesians (but not by frequentists).

Some (frequentist) significance tests are not consistent with the likelihood principle. Bayesians accept the principle which is consistent with their philosophy (perhaps encouraged by the discomfiture of frequentists). "The likelihood approach is compatible with Bayesian statistical inference in the sense that the posterior Bayes distribution for a parameter is, by Bayes's Theorem, found by multiplying the prior distribution by the likelihood function." Frequentists interpret the principle adversely to Bayesians as implying no concern about the reliability of evidence. "The likelihood principle of Bayesian statistics implies that information about the experimental design from which evidence is collected does not enter into the statistical analysis of the data." Many Bayesians (Savage for example) recognize that implication as a vulnerability.

The likelihood principle has become an embarrassment to both major philosophical schools of statistics; It has weakened both rather than favoring either. Its strongest supporters claim that it offers a better foundation for statistics than either of the two schools. "[L]ikelihood looks very good indeed when it is compared with these [Bayesian and frequentist] alternatives." These supporters include statisticians and philosophers of science. The concept needs further development before it can be regarded as a serious challenge to either existing school, but it seems to offer a promising compromise position. While Bayesians acknowledge the importance of likelihood for calculation, they believe that the posterior probability distribution is the proper basis for inference.

Modeling

Inferential statistics is based on models. Much of classical hypothesis testing, for example, was based on the assumed normality of the data. Robust and nonparametric statistics were developed to reduce the dependence on that assumption. Bayesian statistics interprets new observations from the perspective of prior knowledge – assuming a modeled continuity between past and present. The design of experiments assumes some knowledge of those factors to be controlled, varied, randomized and observed. Statisticians are well aware of the difficulties in proving causation (more of a modeling limitation than a mathematical one), saying "correlation does not imply causation".

More complex statistics utilizes more complex models, often with the intent of finding a latent structure underlying a set of variables. As models and data sets have grown in complexity, foundational questions have been raised about the justification of the models and the validity of inferences drawn from them. The range of conflicting opinion expressed about modeling is large.

- Models can be based on scientific theory or on ad-hoc data analysis. The approaches use different methods. There are advocates of each.

- Model complexity is a compromise. The Akaikean information criterion and Bayesian information criterion are two less subjective approaches to achieving that compromise.

- Fundamental reservations have been expressed about even simple regression models used in the social sciences. A long list of assumptions inherent to the validity of a model is typically neither mentioned nor checked. A favorable comparison between observations and model is often considered sufficient.

- Bayesian statistics focuses so tightly on the posterior probability that it ignores the fundamental comparison of observations and model.

- Traditional observation-based models are inadequate to solve many important problems. A much wider range of models, including algorithmic models, must be utilized. "If the model is a poor emulation of nature, the conclusions may be wrong."

- Modeling is often poorly done (the wrong methods are used) and poorly reported.

In the absence of a strong philosophical consensus review of statistical modeling, many statisticians accept the cautionary words of statistician George Box, "All models are wrong, but some are useful."

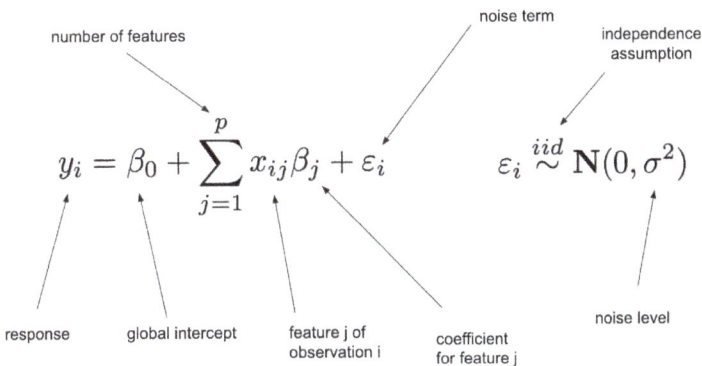

$$y_i = \beta_0 + \sum_{j=1}^{p} x_{ij}\beta_j + \varepsilon_i \qquad \varepsilon_i \overset{iid}{\sim} \mathbf{N}(0, \sigma^2)$$

number of features · noise term · independence assumption · response · global intercept · feature j of observation i · coefficient for feature j · noise level

References

- Yates, Daniel S.; Moore, David S; Starnes, Daren S. (2003). The Practice of Statistics (2nd ed.). New York: Freeman. ISBN 978-0-7167-4773-4.

- Zelterman, Daniel (2010). Applied linear models with SAS ([Online-Ausg.]. ed.). Cambridge: Cambridge University Press. ISBN 9780521761598.

- Cook, R. Dennis; Weisberg, Sanford (1982). Residuals and Influence in Regression. (Repr. ed.). New York: Chapman and Hall. ISBN 041224280X. Retrieved 23 February 2013.

- Weisberg, Sanford (1985). Applied Linear Regression (2nd ed.). New York: Wiley. ISBN 9780471879572. Retrieved 23 February 2013.

- Hazewinkel, Michiel, ed. (2001), "Errors, theory of", Encyclopedia of Mathematics, Springer, ISBN 978-1-55608-010-4

- Lehmann, E. L.; Casella, George (1998). Theory of Point Estimation (2nd ed.). New York: Springer. ISBN 0-387-98502-6. MR 1639875.

- Wackerly, Dennis; Mendenhall, William; Scheaffer, Richard L. (2008). Mathematical Statistics with Applications (7 ed.). Belmont, CA, USA: Thomson Higher Education. ISBN 0-495-38508-5.

- Stigler, Stephen M. (1986). The History of Statistics: The Measurement of Uncertainty Before 1900. Cambridge, MA: Belknap Press of Harvard University Press. ISBN 0-674-40340-1.

- For a good introduction to error-in-variables, please see Fuller, W. A. (1987). Measurement Error Models. John Wiley & Sons. ISBN 0-471-86187-1.

- Hastie, Trevor; Tibshirani, Robert; Friedman, Jerome H. (2009). "The Elements of Statistical Learning" (second ed.). Springer-Verlag. ISBN 978-0-387-84858-7.

- Bühlmann, Peter; van de Geer, Sara (2011). Statistics for High-Dimensional Data: Methods, Theory and Applications. Springer. ISBN 9783642201929.

- B. S. Everitt: The Cambridge Dictionary of Statistics, Cambridge University Press, Cambridge (3rd edition, 2006). ISBN 0-521-69027-7

Methodologies of Statistics

This chapter explains to the reader the methodologies of statistics. The techniques analyzed in the text are descriptive statistics, statistical inference, univariate analysis, etc. They make possible the various types analysis that this field accomplishes. These aspects are of vital importance, as it provides a better understanding of statistics.

Descriptive Statistics

Descriptive statistics is the discipline of quantitatively describing the main features of a collection of information, or the quantitative description itself. Descriptive statistics are distinguished from inferential statistics (or inductive statistics), in that descriptive statistics aim to summarize a sample, rather than use the data to learn about the population that the sample of data is thought to represent. This generally means that descriptive statistics, unlike inferential statistics, are not developed on the basis of probability theory. Even when a data analysis draws its main conclusions using inferential statistics, descriptive statistics are generally also presented. For example in a paper reporting on a study involving human subjects, there typically appears a table giving the overall sample size, sample sizes in important subgroups (e.g., for each treatment or exposure group), and demographic or clinical characteristics such as the average age, the proportion of subjects of each sex, and the proportion of subjects with related comorbidities.

Some measures that are commonly used to describe a data set are measures of central tendency and measures of variability or dispersion. Measures of central tendency include the mean, median and mode, while measures of variability include the standard deviation (or variance), the minimum and maximum values of the variables, kurtosis and skewness.

Use in Statistical Analysis

Descriptive statistics provide simple summaries about the sample and about the observations that have been made. Such summaries may be either quantitative, i.e. summary statistics, or visual, i.e. simple-to-understand graphs. These summaries may either form the basis of the initial description of the data as part of a more extensive statistical analysis, or they may be sufficient in and of themselves for a particular investigation.

For example, the shooting percentage in basketball is a descriptive statistic that summarizes the performance of a player or a team. This number is the number of shots made divided by the number of shots taken. For example, a player who shoots 33% is making approximately one shot in every three. The percentage summarizes or describes multiple discrete events. Consider also the grade point average. This single number describes the general performance of a student across the range of their course experiences.

The use of descriptive and summary statistics has an extensive history and, indeed, the simple tabulation of populations and of economic data was the first way the topic of statistics appeared. More recently, a collection of summarisation techniques has been formulated under the heading of exploratory data analysis: an example of such a technique is the box plot.

In the business world, descriptive statistics provides a useful summary of many types of data. For example, investors and brokers may use a historical account of return behavior by performing empirical and analytical analyses on their investments in order to make better investing decisions in the future.

Univariate Analysis

Univariate analysis involves describing the distribution of a single variable, including its central tendency (including the mean, median, and mode) and dispersion (including the range and quantiles of the data-set, and measures of spread such as the variance and standard deviation). The shape of the distribution may also be described via indices such as skewness and kurtosis. Characteristics of a variable's distribution may also be depicted in graphical or tabular format, including histograms and stem-and-leaf display.

Bivariate Analysis

When a sample consists of more than one variable, descriptive statistics may be used to describe the relationship between pairs of variables. In this case, descriptive statistics include:

- Cross-tabulations and contingency tables

- Graphical representation via scatterplots

- Quantitative measures of dependence

- Descriptions of conditional distributions

The main reason for differentiating univariate and bivariate analysis is that bivariate analysis is not only simple descriptive analysis, but also it describes the relationship between two different variables. Quantitative measures of dependence include correlation (such as Pearson's r when both variables are continuous, or Spearman's rho if one or both are not) and covariance (which reflects the scale variables are measured on). The slope, in regression analysis, also reflects the relationship between variables. The unstandardised slope indicates the unit change in the criterion variable for a one unit change in the predictor. The standardised slope indicates this change in standardised (z-score) units. Highly skewed data are often transformed by taking logarithms. Use of logarithms makes graphs more symmetrical and look more similar to the normal distribution, making them easier to interpret intuitively.

Statistical Inference

Statistical inference is the process of deducing properties of an underlying distribution by analysis of data. Inferential statistical analysis infers properties about a population: this includes testing hypotheses and deriving estimates. The population is assumed to be larger than the observed data set; in other words, the observed data is assumed to be sampled from a larger population.

Inferential statistics can be contrasted with descriptive statistics. Descriptive statistics is solely concerned with properties of the observed data, and does not assume that the data came from a larger population.

Introduction

Statistical inference makes propositions about a population, using data drawn from the population with some form of sampling. Given a hypothesis about a population, for which we wish to draw inferences, statistical inference consists of (firstly) selecting a statistical model of the process that generates the data and (secondly) deducing propositions from the model.

Konishi & Kitagawa state, "The majority of the problems in statistical inference can be considered to be problems related to statistical modeling". Relatedly, Sir David Cox has said, "How [the] translation from subject-matter problem to statistical model is done is often the most critical part of an analysis".

The conclusion of a statistical inference is a statistical proposition.Some common forms of statistical proposition are the following:

- a point estimate, i.e. a particular value that best approximates some parameter of interest;

- an interval estimate, e.g. a confidence interval (or set estimate), i.e. an interval constructed using a dataset drawn from a population so that, under repeated sampling of such datasets, such intervals would contain the true parameter value with the probability at the stated confidence level;

- a credible interval, i.e. a set of values containing, for example, 95% of posterior belief;

- rejection of a hypothesis;

- clustering or classification of data points into groups.

Models and Assumptions

Any statistical inference requires some assumptions. A statistical model is a set of assumptions concerning the generation of the observed data and similar data. Descriptions of statistical models usually emphasize the role of population quantities of interest, about which we wish to draw inference. Descriptive statistics are typically used as a preliminary step before more formal inferences are drawn.

Degree of Models/Assumptions

Statisticians distinguish between three levels of modeling assumptions;

- Fully parametric: The probability distributions describing the data-generation process are assumed to be fully described by a family of probability distributions involving only a finite number of unknown parameters. For example, one may assume that the distribution of population values is truly Normal, with unknown mean and variance, and that datasets are generated by 'simple' random sampling. The family of generalized linear models is a widely used and flexible class of parametric models.

- Non-parametric: The assumptions made about the process generating the data are much less than in parametric statistics and may be minimal. For example, every continuous probability distribution has a median, which may be estimated using the sample median or the Hodges–Lehmann–Sen estimator, which has good properties when the data arise from simple random sampling.

- Semi-parametric: This term typically implies assumptions 'in between' fully and non-parametric approaches. For example, one may assume that a population distribution has a finite mean. Furthermore, one may assume that the mean response level in the population depends in a truly linear manner on some covariate (a parametric assumption) but not make any parametric assumption describing the variance around that mean (i.e. about the presence or possible form of any heteroscedasticity). More generally, semi-parametric models can often be separated into 'structural' and 'random variation' components. One component is treated parametrically and the other non-parametrically. The well-known Cox model is a set of semi-parametric assumptions.

Importance of Valid Models/Assumptions

Whatever level of assumption is made, correctly calibrated inference in general requires these assumptions to be correct; i.e. that the data-generating mechanisms really have been correctly specified.

Incorrect assumptions of 'simple' random sampling can invalidate statistical inference. More complex semi- and fully parametric assumptions are also cause for concern. For example, incorrectly assuming the Cox model can in some cases lead to faulty conclusions. Incorrect assumptions of Normality in the population also invalidates some forms of regression-based inference. The use of any parametric model is viewed skeptically by most experts in sampling human populations: "most sampling statisticians, when they deal with confidence intervals at all, limit themselves to statements about [estimators] based on very large samples, where the central limit theorem ensures that these [estimators] will have distributions that are nearly normal." In particular, a normal distribution "would be a totally unrealistic and catastrophically unwise assumption to make if we were dealing with any kind of economic population." Here, the central limit theorem states that the distribution of the sample mean "for very large samples" is approximately normally distributed, if the distribution is not heavy tailed.

Approximate Distributions

Given the difficulty in specifying exact distributions of sample statistics, many methods have been developed for approximating these.

With finite samples, approximation results measure how close a limiting distribution approaches the statistic's sample distribution: For example, with 10,000 independent samples the normal distribution approximates (to two digits of accuracy) the distribution of the sample mean for many population distributions, by the Berry–Esseen theorem. Yet for many practical purposes, the normal approximation provides a good approximation to the sample-mean's distribution when there are 10 (or more) independent samples, according to simulation studies and statisticians' experience. Following Kolmogorov's work in the 1950s, advanced statistics uses approximation theory and functional analysis to quantify the error of approximation. In this approach, the metric geometry of probability distributions is studied; this approach quantifies approximation error with, for example, the Kullback–Leibler divergence, Bregman divergence, and the Hellinger distance.

With indefinitely large samples, limiting results like the central limit theorem describe the sample statistic's limiting distribution, if one exists. Limiting results are not statements about finite samples, and indeed are irrelevant to finite samples. However, the asymptotic theory of limiting distributions is often invoked for work with finite samples. For example, limiting results are often invoked to justify the generalized method of moments and the use of generalized estimating equations, which are popular in econometrics and biostatistics. The magnitude of the difference between the limiting distribution and the true distribution (formally, the 'error' of the approximation) can be assessed using simulation. The heuristic application of limiting results to finite samples is common practice in many applications, especially with low-dimensional models with log-concave likelihoods (such as with one-parameter exponential families).

Randomization-based Models

For a given dataset that was produced by a randomization design, the randomization distribution of a statistic (under the null-hypothesis) is defined by evaluating the test statistic for all of the plans that could have been generated by the randomization design. In frequentist inference, randomization allows inferences to be based on the randomization distribution rather than a subjective model, and this is important especially in survey sampling and design of experiments. Statistical inference from randomized studies is also more straightforward than many other situations. In Bayesian inference, randomization is also of importance: in survey sampling, use of sampling without replacement ensures the exchangeability of the sample with the population; in randomized experiments, randomization warrants a missing at random assumption for covariate information.

Objective randomization allows properly inductive procedures. Many statisticians prefer randomization-based analysis of data that was generated by well-defined randomization procedures. (However, it is true that in fields of science with developed theoretical knowledge and experimental control, randomized experiments may increase the costs of experimentation without improving the quality of inferences.) Similarly, results from randomized experiments are recommended by leading statistical authorities as allowing inferences with greater reliability than do observational studies of the same phenomena. However, a good observational study may be better than a bad randomized experiment.

The statistical analysis of a randomized experiment may be based on the randomization scheme stated in the experimental protocol and does not need a subjective model.

However, at any time, some hypotheses cannot be tested using objective statistical models, which accurately describe randomized experiments or random samples. In some cases, such randomized studies are uneconomical or unethical.

Model-based Analysis of Randomized Experiments

It is standard practice to refer to a statistical model, often a linear model, when analyzing data from randomized experiments. However, the randomization scheme guides the choice of a statistical model. It is not possible to choose an appropriate model without knowing the randomization scheme. Seriously misleading results can be obtained analyzing data from randomized experiments while ignoring the experimental protocol; common mistakes include forgetting the blocking used in an experiment and confusing repeated measurements on the same experimental unit with independent replicates of the treatment applied to different experimental units.

Paradigms for Inference

Different schools of statistical inference have become established. These schools—or "paradigms"—are not mutually exclusive, and methods that work well under one paradigm often have attractive interpretations under other paradigms.

Bandyopadhyay & Forster describe four paradigms: "(i) classical statistics or error statistics, (ii) Bayesian statistics, (iii) likelihood-based statistics, and (iv) the Akaikean-Information Criterion-based statistics". The classical (or frequentist) paradigm, the Bayesian paradigm, and the AIC-based paradigm are summarized below. The likelihood-based paradigm is essentially a sub-paradigm of the AIC-based paradigm.

Frequentist Inference

This paradigm calibrates the plausibility of propositions by considering (notional) repeated sampling of a population distribution to produce datasets similar to the one at hand. By considering the dataset's characteristics under repeated sampling, the frequentist properties of a statistical proposition can be quantified—although in practice this quantification may be challenging.

Examples of Frequentist Inference

- p-value

- Confidence interval

Frequentist Inference, Objectivity, and Decision Theory

One interpretation of frequentist inference (or classical inference) is that it is applicable only in terms of frequency probability; that is, in terms of repeated sampling from a population. However, the approach of Neyman develops these procedures in terms of pre-experiment probabilities. That is, before undertaking an experiment, one decides on a rule for coming to a conclusion such that the probability of being correct is controlled in a suitable way: such a probability need not have a frequentist or repeated sampling interpretation. In contrast, Bayesian inference works in terms of conditional probabilities (i.e. probabilities conditional on the observed data), compared to the marginal (but conditioned on unknown parameters) probabilities used in the frequentist approach.

The frequentist procedures of significance testing and confidence intervals can be constructed without regard to utility functions. However, some elements of frequentist statistics, such as statistical decision theory, do incorporate utility functions. In particular, frequentist developments of optimal inference (such as minimum-variance unbiased estimators, or uniformly most powerful testing) make use of loss functions, which play the role of (negative) utility functions. Loss functions need not be explicitly stated for statistical theorists to prove that a statistical procedure has an optimality property. However, loss-functions are often useful for stating optimality properties: for example, median-unbiased estimators are optimal under absolute value loss functions, in that they minimize expected loss, and least squares estimators are optimal under squared error loss functions, in that they minimize expected loss.

While statisticians using frequentist inference must choose for themselves the parameters of interest, and the estimators/test statistic to be used, the absence of obviously explicit utilities and prior distributions has helped frequentist procedures to become widely viewed as 'objective'.

Bayesian Inference

The Bayesian calculus describes degrees of belief using the 'language' of probability; beliefs are positive, integrate to one, and obey probability axioms. Bayesian inference uses the available posterior beliefs as the basis for making statistical propositions. There are several different justifications for using the Bayesian approach.

Examples of Bayesian Inference

- Credible interval for interval estimation

- Bayes factors for model comparison

Bayesian Inference, Subjectivity and Decision Theory

Many informal Bayesian inferences are based on "intuitively reasonable" summaries of the posterior. For example, the posterior mean, median and mode, highest posterior density intervals, and Bayes Factors can all be motivated in this way. While a user's utility function need not be stated for this sort of inference, these summaries do all depend (to some extent) on stated prior beliefs, and are generally viewed as subjective conclusions. (Methods of prior construction which do not require external input have been proposed but not yet fully developed.)

Formally, Bayesian inference is calibrated with reference to an explicitly stated utility, or loss function; the 'Bayes rule' is the one which maximizes expected utility, averaged over the posterior uncertainty. Formal Bayesian inference therefore automatically provides optimal decisions in a decision theoretic sense. Given assumptions, data and utility, Bayesian inference can be made for essentially any problem, although not every statistical inference need have a Bayesian interpretation. Analyses which are not formally Bayesian can be (logically) incoherent; a feature of Bayesian procedures which use proper priors (i.e. those integrable to one) is that they are guaranteed to be coherent. Some advocates of Bayesian inference assert that inference *must* take place in this decision-theoretic framework, and that Bayesian inference should not conclude with the evaluation and summarization of posterior beliefs.

AIC-based Inference

Minimum Description Length

The minimum description length (MDL) principle has been developed from ideas in information theory and the theory of Kolmogorov complexity. The (MDL) principle selects statistical models that maximally compress the data; inference proceeds without assuming counterfactual or non-falsifiable "data-generating mechanisms" or probability models for the data, as might be done in frequentist or Bayesian approaches.

However, if a "data generating mechanism" does exist in reality, then according to Shannon's source coding theorem it provides the MDL description of the data, on average and asymptotically. In minimizing description length (or descriptive complexity), MDL estimation is similar to maximum likelihood estimation and maximum a posteriori estimation (using maximum-entropy Bayesian priors). However, MDL avoids assuming that the underlying probability model is known; the MDL principle can also be applied without assumptions that e.g. the data arose from independent sampling.

The MDL principle has been applied in communication-coding theory in information theory, in linear regression, and in data mining.

The evaluation of MDL-based inferential procedures often uses techniques or criteria from computational complexity theory.

Fiducial Inference

Fiducial inference was an approach to statistical inference based on fiducial probability, also known as a "fiducial distribution". In subsequent work, this approach has been called ill-defined, extremely limited in applicability, and even fallacious. However this argument is the same as that which shows that a so-called confidence distribution is not a valid probability distribution and, since this has not invalidated the application of confidence intervals, it does not necessarily invalidate conclusions drawn from fiducial arguments. An attempt was made to reinterpret the early work of Fisher's fiducial argument as a special case of an inference theory using Upper and lower probabilities.

Structural Inference

Developing ideas of Fisher and of Pitman from 1938 to 1939, George A. Barnard developed "structural inference" or "pivotal inference", an approach using invariant probabilities on group families. Barnard reformulated the arguments behind fiducial inference on a restricted class of models on which "fiducial" procedures would be well-defined and useful.

Inference Topics

The topics below are usually included in the area of statistical inference.

1. Statistical assumptions
2. Statistical decision theory
3. Estimation theory
4. Statistical hypothesis testing
5. Revising opinions in statistics
6. Design of experiments, the analysis of variance, and regression
7. Survey sampling
8. Summarizing statistical data

Univariate Analysis

Univariate analysis is perhaps the simplest form of statistical analysis. Like other forms of statistics, it can be inferential or descriptive. The key fact is that only one variable is involved.

Descriptive Methods

Descriptive statistics describe a sample or population. They can be part of exploratory data analysis.

The appropriate statistic depends on the level of measurement. For nominal variables, a frequency table and a listing of the mode(s) is sufficient. For ordinal variables the median can be calculated as a measure of central tendency and the range (and variations of it) as a measure of dispersion. For interval level variables, the arithmetic mean (average) and standard deviation are added to the toolbox and, for ratio level variables, we add the geometric mean and harmonic mean as measures of central tendency and the coefficient of variation as a measure of dispersion.

For interval and ratio level data, further descriptors include the variable's skewness and kurtosis.

Inferential Methods

Inferential methods allow us to infer from a sample to a population. For a nominal variable a one-way chi-square (goodness of fit) test can help determine if our sample matches that of some population. For interval and ratio level data, a one-sample t-test can let us infer whether the mean in our sample matches some proposed number (typically 0). Other available tests of location include the one-sample sign test and Wilcoxon signed rank test.

Bivariate Analysis

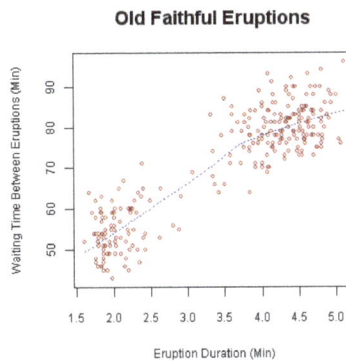

Waiting time between eruptions and the duration of the eruption for the Old Faithful Geyser in Yellowstone National Park, Wyoming, USA. This scatterplot suggests there are generally two "types" of eruptions: short-wait-short-duration, and long-wait-long-duration.

Bivariate analysis is one of the simplest forms of quantitative (statistical) analysis. It involves the analysis of two variables (often denoted as X, Y), for the purpose of determining the empirical relationship between them.

Bivariate analysis can be helpful in testing simple hypotheses of association. Bivariate analysis can help determine to what extent it becomes easier to know and predict a value for one variable (possibly a dependent variable) if we know the value of the other variable (possibly the independent variable) .

Bivariate analysis can be contrasted with univariate analysis in which only one variable is analysed. Like univariate analysis, bivariate analysis can be descriptive or inferential. It is the analysis of the relationship between the two variables. Bivariate analysis is a simple (two variable) special case of multivariate analysis (where multiple relations between multiple variables are examined simultaneously).

There are 2 types, one is inferential and the other is descriptive.

When there is a Dependent Variable

If the dependent variable—the one whose value is determined to some extent by the other, independent variable— is a categorical variable, such as the preferred brand of cereal, then probit or logit regression (or multinomial probit or multinomial logit) can be used. If both variables are ordinal, meaning they are ranked in a sequence as first, second, etc., then a rank correlation coefficient can be computed. If just the dependent variable is ordinal, ordered probit or ordered logit can be used. If the dependent variable is continuous—either interval level or ratio level, such as a temperature scale or an income scale—then simple regression can be used.

If both variables are time series, a particular type of causality known as Granger causality can be tested for, and vector autoregression can be performed to examine the intertemporal linkages between the variables.

When there is not a Dependent Variable

When neither variable can be regarded as dependent on the other, regression is not appropriate but some form of correlation analysis may be.

Graphical Methods

Graphs that are appropriate for bivariate analysis depend on the type of variable. For two continuous variables, a scatterplot is a common graph. When one variable is categorical and the other continuous, a box plot is common and when both are categorical a mosaic plot is common. These graphs are part of descriptive statistics.

Multivariate Statistics

Multivariate statistics is a subdivision of statistics encompassing the simultaneous observation and analysis of more than one outcome variable. The application of multivariate statistics is multivariate analysis.

Multivariate statistics concerns understanding the different aims and background of each of the different forms of multivariate analysis, and how they relate to each other. The practical implementation of multivariate statistics to a particular problem may involve several types of univariate and multivariate analyses in order to understand the relationships between variables and their relevance to the actual problem being studied.

In addition, multivariate statistics is concerned with multivariate probability distributions, in terms of both

- how these can be used to represent the distributions of observed data;

- how they can be used as part of statistical inference, particularly where several different quantities are of interest to the same analysis.

Certain types of problem involving multivariate data, for example simple linear regression and multiple regression, are *not* usually considered as special cases of multivariate statistics because the analysis is dealt with by considering the (univariate) conditional distribution of a single outcome variable given the other variables.

Types of Analysis

There are many different models, each with its own type of analysis:

1. Multivariate analysis of variance (MANOVA) extends the analysis of variance to cover cases where there is more than one dependent variable to be analyzed simultaneously; .

2. Multivariate regression attempts to determine a formula that can describe how elements in a vector of variables respond simultaneously to changes in others. For linear relations, regression analyses here are based on forms of the general linear model. Note that Multivariate regression is distinct from Multivariable regression, which has only one dependent variable.

3. Principal components analysis (PCA) creates a new set of orthogonal variables that contain the same information as the original set. It rotates the axes of variation to give a new set of orthogonal axes, ordered so that they summarize decreasing proportions of the variation.

4. Factor analysis is similar to PCA but allows the user to extract a specified number of synthetic variables, fewer than the original set, leaving the remaining unexplained variation as error. The extracted variables are known as latent variables or factors; each one may be supposed to account for covariation in a group of observed variables.

5. Canonical correlation analysis finds linear relationships among two sets of variables; it is the generalised (i.e. canonical) version of bivariate correlation.

6. Redundancy analysis (RDA) is similar to canonical correlation analysis but allows the user to derive a specified number of synthetic variables from one set of (independent) variables that explain as much variance as possible in another (independent) set. It is a multivariate analogue of regression.

7. Correspondence analysis (CA), or reciprocal averaging, finds (like PCA) a set of synthetic variables that summarise the original set. The underlying model assumes chi-squared dissimilarities among records (cases).

8. Canonical (or "constrained") correspondence analysis (CCA) for summarising the joint variation in two sets of variables (like redundancy analysis); combination of correspondence analysis and multivariate regression analysis. The underlying model assumes chi-squared dissimilarities among records (cases).

9. Multidimensional scaling comprises various algorithms to determine a set of synthetic variables that best represent the pairwise distances between records. The original method is principal coordinates analysis (PCoA; based on PCA).

10. Discriminant analysis, or canonical variate analysis, attempts to establish whether a set of variables can be used to distinguish between two or more groups of cases.

11. Linear discriminant analysis (LDA) computes a linear predictor from two sets of normally distributed data to allow for classification of new observations.

12. Clustering systems assign objects into groups (called clusters) so that objects (cases) from the same cluster are more similar to each other than objects from different clusters.

13. Recursive partitioning creates a decision tree that attempts to correctly classify members of the population based on a dichotomous dependent variable.

14. Artificial neural networks extend regression and clustering methods to non-linear multivariate models.

15. Statistical graphics such as tours, parallel coordinate plots, scatterplot matrices can be used to explore multivariate data.

16. Simultaneous equations models involve more than one regression equation, with different dependent variables, estimated together.

17. Vector autoregression involves simultaneous regressions of various time series variables on their own and each other's lagged values.

Important Probability Distributions

There is a set of probability distributions used in multivariate analyses that play a similar role to the corresponding set of distributions that are used in univariate analysis when the normal distribution is appropriate to a dataset. These multivariate distributions are:

- Multivariate normal distribution

- Wishart distribution

- Multivariate Student-t distribution.

The Inverse-Wishart distribution is important in Bayesian inference, for example in Bayesian multivariate linear regression. Additionally, Hotelling's T-squared distribution is a multivariate distribution, generalising Student's t-distribution, that is used in multivariate hypothesis testing.

History

Anderson's 1958 textbook, *An Introduction to Multivariate Analysis*, educated a generation of theorists and applied statisticians; Anderson's book emphasizes hypothesis testing via likelihood ratio tests and the properties of power functions: Admissibility, unbiasedness and monotonicity.

Software and Tools

There are an enormous number of software packages and other tools for multivariate analysis, including:

- High-D
- JMP (statistical software)
- MiniTab
- Calc
- PLS_Toolbox / Solo (Eigenvector Research)
- PSPP
- SAS (software)
- SciPy for Python
- SPSS
- Stata
- STATISTICA
- TMVA - Toolkit for Multivariate Data Analysis in ROOT
- The Unscrambler
- SmartPLS - Next Generation Path Modeling
- MATLAB
- Eviews
- Prosensus ProMV
- Umetrics SIMCA

Structured Data Analysis (Statistics)

Structured data analysis is the statistical data analysis of structured data. This can arise either in the form of an *a priori* structure such as multiple-choice questionnaires or in situations with the need to search for structure that fits the given data, either exactly or approximately. This structure can then be used for making comparisons, predictions, manipulations etc.

Types of Structured Data Analysis

- Algebraic data analysis
- Bayesian analysis
- Cluster analysis
- Combinatorial data analysis
- Formal concept analysis
- Functional data analysis
- Geometric data analysis
- Regression analysis
- Shape analysis
- Topological data analysis
- Tree structured data analysis

References

- Babbie, Earl R. (2009). The Practice of Social Research (12th ed.). Wadsworth. pp. 436–440. ISBN 0-495-59841-0.

- Nick, Todd G. (2007). "Descriptive Statistics". Topics in Biostatistics. Methods in Molecular Biology 404. New York: Springer. pp. 33–52. doi:10.1007/978-1-59745-530-5_3. ISBN 978-1-58829-531-6

- Bickel, Peter J.; Doksum, Kjell A. (2001). Mathematical statistics: Basic and selected topics. 1 (Second (updated printing 2007) ed.). Prentice Hall. ISBN 0-13-850363-X. MR 443141.

- Freedman, D. A. (2009). Statistical models: Theory and practice (revised ed.). Cambridge University Press. pp. xiv+442 pp. ISBN 978-0-521-74385-3. MR 2489600.

- Hinkelmann, Klaus; Kempthorne, Oscar (2008). Introduction to Experimental Design (Second ed.). Wiley. ISBN 978-0-471-72756-9.

- Le Cam, Lucian. (1986) Asymptotic Methods of Statistical Decision Theory, Springer. ISBN 0-387-96307-3

- Peirce, C. S. (1883), "A Theory of Probable Inference", Studies in Logic, pp. 126-181, Little, Brown, and Company. (Reprinted 1983, John Benjamins Publishing Company, ISBN 90-272-3271-7)

- Pfanzagl, Johann; with the assistance of R. Hamböker (1994). Parametric Statistical Theory. Berlin: Walter de Gruyter. ISBN 3-11-013863-8. MR 1291393.

- Rissanen, Jorma (1989). Stochastic Complexity in Statistical Inquiry. Series in computer science. 15. Singapore: World Scientific. ISBN 9971-5-0859-1. MR 1082556.

- Traub, Joseph F.; Wasilkowski, G. W.; Wozniakowski, H. (1988). Information-Based Complexity. Academic Press. ISBN 0-12-697545-0.

- Everitt, Brian (1998). The Cambridge Dictionary of Statistics. Cambridge, UK New York: Cambridge University Press. ISBN 0521593468.

Measures of Central Tendency

A central tendency is a concise form of data that represents whole categories of data. Central tendency can be measured in averages. The measures of the subject matter are mean, median and mode and they indicate a different property of the central tendency. The topics discussed in the chapter are of great importance to broaden the existing knowledge on statistics.

Mean

In mathematics, mean has several different definitions depending on the context.

In probability and statistics, mean and expected value are used synonymously to refer to one measure of the central tendency either of a probability distribution or of the random variable characterized by that distribution. In the case of a discrete probability distribution of a random variable X, the mean is equal to the sum over every possible value weighted by the probability of that value; that is, it is computed by taking the product of each possible value x of X and its probability $P(x)$, and then adding all these products together, giving $\mu = \sum xP(x)$. An analogous formula applies to the case of a continuous probability distribution. Not every probability distribution has a defined mean. Moreover, for some distributions the mean is infinite: for example, when the probability of the value 2^n is $\frac{1}{2^n}$ for n = 1, 2, 3, ...

For a data set, the terms arithmetic mean, mathematical expectation, and sometimes average are used synonymously to refer to a central value of a discrete set of numbers: specifically, the sum of the values divided by the number of values. The arithmetic mean of a set of numbers $x_1, x_2, ..., x_n$ is typically denoted by $\bar{x}$, pronounced "x bar". If the data set were based on a series of observations obtained by sampling from a statistical population, the arithmetic mean is termed the sample mean (denoted $\bar{x}$) to distinguish it from the population mean (denoted μ or μ_x).

For a finite population, the population mean of a property is equal to the arithmetic mean of the given property while considering every member of the population. For example, the population mean height is equal to the sum of the heights of every individual divided by the total number of individuals. The sample mean may differ from the population mean, especially for small samples. The law of large numbers dictates that the larger the size of the sample, the more likely it is that the sample mean will be close to the population mean.

Outside of probability and statistics, a wide range of other notions of "mean" are often used in geometry and analysis; examples are given below.

Types of Mean

Arithmetic Mean (Am)

The *arithmetic mean* (or simply "mean") of a sample $x_1, x_2, \ldots, x_n$, usually denoted by $\overline{x}$, is the sum of the sampled values divided by the number of items in the sample:

$$\overline{x} = \frac{x_1 + x_2 + \cdots + x_n}{n}$$

For example, the arithmetic mean of five values: 4, 36, 45, 50, 75 is

$$\frac{4 + 36 + 45 + 50 + 75}{5} = \frac{210}{5} = 42.$$

Geometric Mean (GM)

The geometric mean is an average that is useful for sets of positive numbers that are interpreted according to their product and not their sum (as is the case with the arithmetic mean) e.g. rates of growth.

$$\overline{x} = \left(\prod_{i=1}^{n} x_i \right)^{\frac{1}{n}}$$

For example, the geometric mean of five values: 4, 36, 45, 50, 75 is:

$$(4 \times 36 \times 45 \times 50 \times 75)^{1/5} = \sqrt[5]{24\,300\,000} = 30.$$

Harmonic Mean (HM)

The harmonic mean is an average which is useful for sets of numbers which are defined in relation to some unit, for example speed (distance per unit of time).

$$\overline{x} = n \cdot \left(\sum_{i=1}^{n} \frac{1}{x_i} \right)^{-1}$$

For example, the harmonic mean of the five values: 4, 36, 45, 50, 75 is

$$\frac{5}{\frac{1}{4} + \frac{1}{36} + \frac{1}{45} + \frac{1}{50} + \frac{1}{75}} = \frac{5}{\frac{1}{3}} = 15.$$

Relationship between AM, GM, and HM

AM, GM, and HM satisfy these inequalities:

$$AM \geq GM \geq HM$$

Equality holds if and only if all the elements of the given sample are equal.

Statistical Location

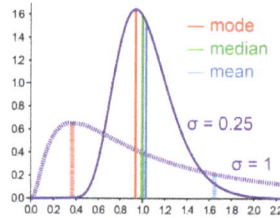

Comparison of the arithmetic mean, median and mode of two skewed (log-normal) distributions.

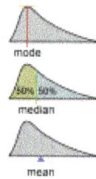

Geometric visualisation of the mode, median and mean of an arbitrary probability density function.

In descriptive statistics, the mean may be confused with the median, mode or mid-range, as any of these may be called an "average" (more formally, a measure of central tendency). The mean of a set of observations is the arithmetic average of the values; however, for skewed distributions, the mean is not necessarily the same as the middle value (median), or the most likely value (mode). For example, mean income is typically skewed upwards by a small number of people with very large incomes, so that the majority have an income lower than the mean. By contrast, the median income is the level at which half the population is below and half is above. The mode income is the most likely income, and favors the larger number of people with lower incomes. While the median and mode are often more intuitive measures for such skewed data, many skewed distributions are in fact best described by their mean, including the exponential and Poisson distributions.

Mean of a Probability Distribution

The mean of a probability distribution is the long-run arithmetic average value of a random variable having that distribution. In this context, it is also known as the expected value. For a discrete probability distribution, the mean is given by $\sum xP(x)$, where the sum is taken over all possible values of the random variable and $P(x)$ is the probability mass function. For a continuous distribution, the mean is $\int_{-\infty}^{\infty} x f(x) dx$, where $f(x)$ is the probability density function. In all cases, including those in which the distribution is neither discrete nor continuous, the mean is the Lebesgue integral of the random variable with respect to its probability measure. The mean need not exist or be finite; for some probability distributions the mean is infinite ($+\infty$ or $-\infty$), while others have no mean.

Generalized Means

Power Mean

The generalized mean, also known as the power mean or Hölder mean, is an abstraction of the quadratic, arithmetic, geometric and harmonic means. It is defined for a set of n positive numbers x_i by

$$\bar{x}(m) = \left(\frac{1}{n} \cdot \sum_{i=1}^{n} x_i^m \right)^{\frac{1}{m}}$$

By choosing different values for the parameter m, the following types of means are obtained:

$m \to \infty$	maximum of x_i
$m = 2$	quadratic mean
$m = 1$	arithmetic mean
$m \to 0$	geometric mean
$m = -1$	harmonic mean
$m \to -\infty$	minimum of x_i

f-mean

This can be generalized further as the generalized f-mean

$$\bar{x} = f^{-1}\left(\frac{1}{n} \cdot \sum_{i=1}^{n} f(x_i) \right)$$

and again a suitable choice of an invertible f will give

$$f(x) \quad x \qquad\qquad \text{arithmetic mean,}$$

$$f(x) = \frac{1}{x} \qquad\qquad \text{harmonic mean,}$$

$$f(x) = x^m \qquad\qquad \text{power mean,}$$

$$f(x) = \ln x \qquad\qquad \text{geometric mean.}$$

Weighted Arithmetic Mean

The weighted arithmetic mean (or weighted average) is used if one wants to combine average values from samples of the same population with different sample sizes:

$$\overline{x} = \frac{\sum_{i=1}^{n} w_i \cdot x_i}{\sum_{i=1}^{n} w_i}.$$

The weights w_i represent the sizes of the different samples. In other applications they represent a measure for the reliability of the influence upon the mean by the respective values.

Truncated Mean

Sometimes a set of numbers might contain outliers, i.e., data values which are much lower or much higher than the others. Often, outliers are erroneous data caused by artifacts. In this case, one can use a truncated mean. It involves discarding given parts of the data at the top or the bottom end, typically an equal amount at each end, and then taking the arithmetic mean of the remaining data. The number of values removed is indicated as a percentage of total number of values.

Interquartile Mean

The interquartile mean is a specific example of a truncated mean. It is simply the arithmetic mean after removing the lowest and the highest quarter of values.

$$\overline{x} = \frac{2}{n} \sum_{i=(n/4)+1}^{3n/4} x_i$$

assuming the values have been ordered, so is simply a specific example of a weighted mean for a specific set of weights.

Mean of a Function

In some circumstances mathematicians may calculate a mean of an infinite (even an uncountable) set of values. This can happen when calculating the mean value y_{ave} of a function $f(x)$. Intuitively this can be thought of as calculating the area under a section of a curve and then dividing by the length of that section. This can be done crudely by counting squares on graph paper or more precisely by integration. The integration formula is written as:

$$y_{ave}(a,b) = \frac{\int_a^b f(x)dx}{b-a}$$

Care must be taken to make sure that the integral converges. But the mean may be finite even if the function itself tends to infinity at some points.

Mean of Angles and Cyclical Quantities

Angles, times of day, and other cyclical quantities require modular arithmetic to add and otherwise combine numbers. In all these situations, there will not be a unique mean. For example, the times an hour before and after midnight are equidistant to both midnight and noon. It is also possible that no mean exists. Consider a color wheel -- there is no mean to the set of all colors. In these situations, you must decide which mean is most useful. You can do this by adjusting the values before averaging, or by using a specialized approach for the mean of circular quantities.

Fréchet Mean

The Fréchet mean gives a manner for determining the "center" of a mass distribution on a surface or, more generally, Riemannian manifold. Unlike many other means, the Fréchet mean is defined on a space whose elements cannot necessarily be added together or multiplied by scalars. It is sometimes also known as the Karcher mean (named after Hermann Karcher).

Distribution of the Sample Mean

The arithmetic mean of a population, or population mean, is denoted μ. The sample mean (the arithmetic mean of a sample of values drawn from the population) makes a good estimator of the population mean, as its expected value is equal to the population mean (that is, it is an unbiased estimator). The sample mean is a random variable, not a constant, since its calculated value will randomly differ depending on which members of the population are sampled, and consequently it will have its own distribution. For a random sample of n observations from a normally distributed population, the sample mean distribution is normally distributed with mean and variance as follows:

$$\bar{x} \sim N\left\{\mu, \frac{\sigma^2}{n}\right\}.$$

Often, since the population *variance* is an unknown parameter, it is estimated by the mean sum of squares; when this estimated value is used, the distribution of the sample mean is no longer a normal distribution but rather a Student's t distribution with $n - 1$ degrees of freedom.

Arithmetic mean

In mathematics and statistics, the arithmetic mean, or simply the mean or average when the context is clear, is the sum of a collection of numbers divided by the number of numbers in the collection. The collection is often a set of results of an experiment, or a set of results from a survey. The term "arithmetic mean" is preferred in some contexts in mathematics and statistics because it helps distinguish it from other means, such as the geometric mean and the harmonic mean.

In addition to mathematics and statistics, the arithmetic mean is used frequently in fields such as economics, sociology, and history, and it is used in almost every academic field to some extent. For example, per capita income is the arithmetic average income of a nation's population.

While the arithmetic mean is often used to report central tendencies, it is not a robust statistic, meaning that it is greatly influenced by outliers (values that are very much larger or smaller than most of the values). Notably, for skewed distributions, such as the distribution of income for which a few people's incomes are substantially greater than most people's, the arithmetic mean may not accord with one's notion of "middle", and robust statistics, such as the median, may be a better description of central tendency.

In a more obscure usage, any sequence of values that form an arithmetic sequence between two numbers x and y can be called "arithmetic means between x and y."

Definition

The arithmetic mean (or mean or average) is the most commonly used and readily understood measure of central tendency. In statistics, the term average refers to any of the measures of central tendency. The arithmetic mean is defined as being equal to the sum of the numerical values of each and every observation divided by the total number of observations. Symbolically, if we have a data set containing the values $a_1, \ldots, a_n$. The arithmetic mean A is defined by the formula

$$A = \frac{1}{n}\sum_{i=1}^{n} a_i.$$

For example, let us consider the monthly salary of 10 employees of a firm: 2500, 2700, 2400, 2300, 2550, 2650, 2750, 2450, 2600, 2400. The arithmetic mean is

$$\frac{2500+2700+2400+2300+2550+2650+2750+2450+2600+2400}{10} = 2530.$$

If the data set is a statistical population (i.e., consists of every possible observation and not just a subset of them), then the mean of that population is called the population mean. If the data set is a statistical sample (a subset of the population), we call the statistic resulting from this calculation a sample mean.

The arithmetic mean of a variable is often denoted by a bar, for example as in $^-$ (read *x bar*), which is the mean of the n values $x_1, x_2, \ldots, x_n$..

Motivating Properties

The arithmetic mean has several properties that make it useful, especially as a measure of central tendency. These include:

- If numbers $x_1, \ldots, x_n$ have mean $\overline{x}$, then $(x_1 - \overline{x}) + \cdots + (x_n - \overline{x}) = 0$.. Since $x_i - \overline{x}$ is the distance from a given number to the mean, one way to interpret this property is as saying that the numbers to the left of the mean are balanced by the numbers to the right of the mean. The mean is the only single number for which the residuals (deviations from the estimate) sum to zero.

- If it is required to use a single number as a "typical" value for a set of known numbers $x_1, \ldots, x_n$, then the arithmetic mean of the numbers does this best, in the sense of minimizing the sum of squared deviations from the typical value: the sum of $(x_i - \overline{x})^2$.. (It follows that the sample mean is also the best single predictor in the sense of having the lowest root mean squared error.) If the arithmetic mean of a population of numbers is desired, then the estimate of it that is unbiased is the arithmetic mean of a sample drawn from the population.

Contrast with Median

The arithmetic mean may be contrasted with the median. The median is defined such that half the values are larger than, and half are smaller than, the median. If elements in the sample data increase arithmetically, when placed in some order, then the median and arithmetic average are equal. For example, consider the data sample $1, 2, 3, 4$.. The average is 2.5,, as is the median. However, when we consider a sample that cannot be arranged so as to increase arithmetically, such as $1, 2, 4, 8, 16$, the median and arithmetic average can differ significantly. In this case, the arithmetic average is 6.2 and the median is 4. In general, the average value can vary significantly from most values in the sample, and can be larger or smaller than most of them.

There are applications of this phenomenon in many fields. For example, since the 1980s, the median income in the United States has increased more slowly than the arithmetic average of income.

Generalizations

Weighted Average

A weighted average, or weighted mean, is an average in which some data points count more strongly than others, in that they are given more weight in the calculation. For example, the arithmetic mean of 3 and 5 is $\frac{(3+5)}{2} = 4$, or equivalently $\left(\frac{1}{2} \cdot 3\right) + \left(\frac{1}{2} \cdot 5\right) = 4$. In contrast, a *weighted* mean in which the first number receives, for example, twice as much weight as the second (perhaps because it is assumed to appear twice as often in the general population from which these numbers were sampled) would be calculated as $\left(\frac{2}{3} \cdot 3\right) + \left(\frac{1}{3} \cdot 5\right) = \frac{11}{3}$. Here the weights, which necessarily sum to the value one, are $(2/3)$ and $(1/3)$, the former being twice the latter. Note that the arithmetic mean (sometimes called the "unweighted average" or "equally weighted average") can be interpreted as a special case of a weighted average in which all the weights are equal to each other (equal to $\frac{1}{2}$ in the above example, and equal to $\frac{1}{n}$ in a situation with n numbers being averaged).

Continuous Probability Distributions

When a population of numbers, and any sample of data from it, could take on any of a continuous range of numbers, instead of for example just integers, then the probability of a number falling into one range of possible values could differ from the probability of falling into a different range of possible values, even if the lengths of both ranges are the same. In such a case, the set of probabilities can be described using a continuous probability distribution. The analog of a weighted average in this context, in which there are an infinitude of possibilities for the precise value of the variable, is called the *mean of the probability distribution*. The most widely encountered probability distribution is called the normal distribution; it has the property that all measures of its central tendency, including not just the mean but also the aforementioned median and the mode, are equal to each other. This property does not hold however, in the cases of a great many probability distributions, such as the lognormal distribution illustrated here.

Angles

Particular care must be taken when using cyclic data, such as phases or angles. Naïvely taking the arithmetic mean of 1° and 359° yields a result of 180°. This is incorrect for two reasons:

- Firstly, angle measurements are only defined up to an additive constant of 360° (or 2π, if measuring in radians). Thus one could as easily call these 1° and −1°, or 361° and 719°, each of which gives a different average.

- Secondly, in this situation, 0° (equivalently, 360°) is geometrically a better *average* value: there is lower dispersion about it (the points are both 1° from it, and 179° from 180°, the putative average).

In general application, such an oversight will lead to the average value artificially moving towards the middle of the numerical range. A solution to this problem is to use the optimization formulation (viz., define the mean as the central point: the point about which one has the lowest dispersion), and redefine the difference as a modular distance (i.e., the distance on the circle: so the modular distance between 1° and 359° is 2°, not 358°).

Geometric mean

In mathematics, the geometric mean is a type of mean or average, which indicates the central tendency or typical value of a set of numbers by using the product of their values (as opposed to the arithmetic mean which uses their sum). The geometric mean is defined as the nth root of the product of n numbers, i.e., for a set of numbers $x_1, x_2, ..., x_n$, the geometric mean is defined as

$$\left(\prod_{i=1}^{n} x_i\right)^{\frac{1}{n}} = \sqrt[n]{x_1 x_2 \cdots x_n}.$$

For instance, the geometric mean of two numbers, say 2 and 8, is just the square root of their product, that is, $\sqrt{2 \cdot 8} = 4..$ As another example, the geometric mean of the three numbers 4, 1, and 1/32 is the cube root of their product (1/8), which is 1/2, that is, $\sqrt[3]{4 \cdot 1 \cdot 1/32} = 1/2$.

A geometric mean is often used when comparing different items—finding a single "figure of merit" for these items—when each item has multiple properties that have different numeric ranges. For example, the geometric mean can give a meaningful "average" to compare two companies which are each rated at 0 to 5 for their environmental sustainability, and are rated at 0 to 100 for their financial viability. If an arithmetic mean were used instead of a geometric mean, the financial viability is given more weight because its numeric range is larger—so a small percentage change in the financial rating (e.g. going from 80 to 90) makes a much larger difference in the arithmetic mean than a large percentage change in environmental sustainability (e.g. going from 2 to 5). The use of a geometric mean "normalizes" the ranges being averaged, so that no range dominates the weighting, and a given percentage change in any of the properties has the same effect on the geometric mean. So, a 20% change in environmental sustainability from 4 to 4.8 has the same effect on the geometric mean as a 20% change in financial viability from 60 to 72.

The geometric mean can be understood in terms of geometry. The geometric mean of two numbers, a and b, is the length of one side of a square whose area is equal to the area of a rectangle with sides of lengths a and b. Similarly, the geometric mean of three numbers, a, b, and c, is the length of one edge of a cube whose volume is the same as that of a cuboid with sides whose lengths are equal to the three given numbers.

The geometric mean applies only to numbers of the same sign. It is also often used for a set of numbers whose values are meant to be multiplied together or are exponential in nature, such as data on the growth of the human population or interest rates of a financial investment.

The geometric mean is also one of the three classical Pythagorean means, together with the aforementioned arithmetic mean and the harmonic mean. For all positive data sets containing at least one pair of unequal values, the harmonic mean is always the least of the three means, while the arithmetic mean is always the greatest of the three and the geometric mean is always in between.

Calculation

The geometric mean of a data set $\{a_1, a_2, \ldots, a_n\}$ is given by:

$$\left(\prod_{i=1}^{n} a_i \right)^{\frac{1}{n}} = \sqrt[n]{a_1 a_2 \cdots a_n}.$$

The above figure uses capital pi notation to show a series of multiplications. In plain English, each side of the equal sign shows that a set of values is multiplied in succession (the number of values is represented by "n") to give a total product of the set, and then the nth root of the total product is taken to give the geometric mean of the original set. For example, in a set of four numbers [1, 2, 3, 4], the product of 1x2x3x4 is 24, and the geometric mean is the fourth root of 24, or ≈ 2.213. Note that the exponent $1/n$ on the left side is equivalent to the taking nth root. For example, $24^{1/4} = \sqrt[4]{24}$.

The geometric mean of a data set is less than the data set's arithmetic mean unless all members of the data set are equal, in which case the geometric and arithmetic means are equal. This allows the definition of the arithmetic-geometric mean, a mixture of the two which always lies in between.

The geometric mean is also the arithmetic-harmonic mean in the sense that if two sequences (a_n) and (h_n) are defined:

$$a_{n+1} = \frac{a_n + h_n}{2}, \quad a_0 = x$$

and

$$h_{n+1} = \frac{2}{\dfrac{1}{a_n} + \dfrac{1}{h_n}}, \quad h_0 = y$$

where h_{n+1} is the harmonic mean of the previous values of the two sequences, then a_n and h_n will converge to the geometric mean of x and y.

This can be seen easily from the fact that the sequences do converge to a common limit (which can be shown by Bolzano–Weierstrass theorem) and the fact that geometric mean is preserved:

$$\sqrt{a_i h_i} = \sqrt{\frac{a_i + h_i}{\dfrac{a_i + h_i}{h_i a_i}}} = \sqrt{\frac{a_i + h_i}{\dfrac{1}{a_i} + \dfrac{1}{h_i}}} = \sqrt{a_{i+1} h_{i+1}}$$

Replacing the arithmetic and harmonic mean by a pair of generalized means of opposite, finite exponents yields the same result.

Relationship with Logarithms

The geometric mean can also be expressed as the arithmetic means of logarithms, By using logarithmic identities to transform the formula, the multiplications can be expressed as a sum and the power as a multiplication:

When $a_1, a_2, \ldots, a_n > 0$

$$\left(\prod_{i=1}^{n} a_i \right)^{\frac{1}{n}} = \exp\left[\frac{1}{n} \sum_{i=1}^{n} \ln a_i \right]$$

if $\exists a_j < 0$ then

$$\left(\prod_{i=1}^{n} a_i\right)^{\frac{1}{n}} = (-1)^{\frac{m}{n}} \exp\left[\frac{1}{n}\sum_{i=1}^{n} \ln|a_i|\right]$$

Where m is the number of negative numbers.

This is sometimes called the log-average. It is simply computing the arithmetic mean of the logarithm-transformed values of a_i (i.e., the arithmetic mean on the log scale) and then using the exponentiation to return the computation to the original scale, i.e., it is the generalised f-mean with $f(x) = \log x.$. For example, the geometric mean of 2 and 8 can be calculated as the following, where b is any base of a logarithm (commonly 2, e or 10):

$$b^{(\log_b(2)+\log_b(8))/2} = 4$$

The log form of the geometric mean is generally the preferred alternative for implementation in computer languages because calculating the product of many numbers can lead to an arithmetic overflow or arithmetic underflow. This is less likely to occur with the sum of the logarithms for each number.

Relationship with Arithmetic Mean and Mean-preserving Spread

If a set of non-identical numbers is subjected to a mean-preserving spread — that is, two or more elements of the set are "spread apart" from each other while leaving the arithmetic mean unchanged — then the geometric mean always decreases.

Computation in Constant Time

In cases where the geometric mean is being used to determine the average growth rate of some quantity, and the initial and final values a_0 and a_n of that quantity are known, the product of the measured growth rate at every step need not be taken. Instead, the geometric mean is simply

$$(\frac{a_n}{a_0})^{\frac{1}{n}},$$

where n is the number of steps from the initial to final state.

If the values are $a_0,\ldots,a_n$, then the growth rate between measurement a_k and a_{k+1} is a_{k+1}/a_k. The geometric mean of these growth rates is just

$$\left(\frac{a_1}{a_0}\frac{a_2}{a_1}\cdots\frac{a_n}{a_{n-1}}\right)^{\frac{1}{n}} = \left(\frac{a_n}{a_0}\right)^{\frac{1}{n}}$$

Properties

The fundamental property of the geometric mean, which can be proven to be false for any other mean, is

$$GM\left(\frac{X_i}{Y_i}\right) = \frac{GM(X_i)}{GM(Y_i)}$$

This makes the geometric mean the only correct mean when averaging *normalized* results; that is, results that are presented as ratios to reference values. This is the case when presenting computer performance with respect to a reference computer, or when computing a single average index from several heterogeneous sources (for example, life expectancy, education years, and infant mortality). In this scenario, using the arithmetic or harmonic mean would change the ranking of the results depending on what is used as a reference. For example, take the following comparison of execution time of computer programs:

	Computer A	Computer B	Computer C
Program 1	1	10	20
Program 2	1000	100	20
Arithmetic mean	500.5	55	20
Geometric mean	31.622 ...	31.622 ...	20
Harmonic mean	1.998 ...	18.182 ...	20

The arithmetic and geometric means "agree" that computer C is the fastest. However, by presenting appropriately normalized values *and* using the arithmetic mean, we can show either of the other two computers to be the fastest. Normalizing by A's result gives A as the fastest computer according to the arithmetic mean:

	Computer A	Computer B	Computer C
Program 1	1	10	20
Program 2	1	0.1	0.02
Arithmetic mean	1	5.05	10.01
Geometric mean	1	1	0.632 ...
Harmonic mean	1	0.198 ...	0.039 ...

while normalizing by B's result gives B as the fastest computer according to the arithmetic mean but A as the fastest according to the harmonic mean:

	Computer A	Computer B	Computer C
Program 1	0.1	1	2

Program 2	10	1	0.2
Arithmetic mean	5.05	1	1.1
Geometric mean	1	1	0.632
Harmonic mean	0.198 . . .	1	0.363 . . .

and normalizing by C's result gives C as the fastest computer according to the arithmetic mean but A as the fastest according to the harmonic mean:

	Computer A	Computer B	Computer C
Program 1	0.05	0.5	1
Program 2	50	5	1
Arithmetic mean	25.025	2.75	1
Geometric mean	1.581 . . .	1.581 . . .	1
Harmonic mean	0.099 . . .	0.909 . . .	1

In all cases, the ranking given by the geometric mean stays the same as the one obtained with unnormalized values.

However, this reasoning has been questioned. Giving consistent results is not always equal to giving the correct results. In general, it is more rigorous to assign weights to each of the programs, calculate the average weighted execution time (using the arithmetic mean), and then normalize that result to one of the computers. The three tables above just give a different weight to each of the programs, explaining the inconsistent results of the arithmetic and harmonic means (the first table gives equal weight to both programs, the second gives a weight of 1/1000 to the second program, and the third gives a weight of 1/100 to the second program and 1/10 to the first one). The use of the geometric mean for aggregating performance numbers should be avoided if possible, because multiplying execution times has no physical meaning, in contrast to adding times as in the arithmetic mean. Metrics that are inversely proportional to time (speedup, IPC) should be averaged using the harmonic mean.

Applications

Proportional Growth

The geometric mean is more appropriate than the arithmetic mean for describing proportional growth, both exponential growth (constant proportional growth) and varying growth; in business the geometric mean of growth rates is known as the compound annual growth rate (CAGR). The geometric mean of growth over periods yields the equivalent constant growth rate that would yield the same final amount.

Suppose an orange tree yields 100 oranges one year and then 180, 210 and 300 the following years, so the growth is 80%, 16.6666% and 42.8571% for each year respectively. Using the arithmetic mean calculates a (linear) average growth of 46.5079% (80% + 16.6666% + 42.8571% divided by 3). However, if we start with 100 oranges and let it grow 46.5079% each year, the result is 314 oranges, not 300, so the linear average *overstates* the year-on-year growth.

Instead, we can use the geometric mean. Growing with 80% corresponds to multiplying with 1.80, so we take the geometric mean of 1.80, 1.166666 and 1.428571, i.e. $\sqrt[3]{1.80 \times 1.166666 \times 1.428571} = 1.442249$; thus the "average" growth per year is 44.2249%. If we start with 100 oranges and let the number grow with 44.2249% each year, the result is 300 oranges.

Applications in the Social Sciences

Although the geometric mean has been relatively rare in computing social statistics, starting from 2010 the United Nations Human Development Index did switch to this mode of calculation, on the grounds that it better reflected the non-substitutable nature of the statistics being compiled and compared:

> *The geometric mean decreases the level of substitutability between dimensions [being compared] and at the same time ensures that a 1 percent decline in say life expectancy at birth has the same impact on the HDI as a 1 percent decline in education or income. Thus, as a basis for comparisons of achievements, this method is also more respectful of the intrinsic differences across the dimensions than a simple average.*

Note that not all values used to compute the HDI (Human Development Index) are normalized; some of them instead have the form $(X - X_{min}) / (X_{norm} - X_{min})$. This makes the choice of the geometric mean less obvious than one would expect from the "Properties" section above.

Aspect Ratios

Equal area comparison of the aspect ratios used by Kerns Powers to derive the SMPTE 16:9 standard. TV 4:3/1.33 in red, 1.66 in orange, 16:9/1.77 in blue, 1.85 in yellow, Panavision/2.2 in mauve and CinemaScope/2.35 in purple.

The geometric mean has been used in choosing a compromise aspect ratio in film and video: given two aspect ratios, the geometric mean of them provides a compromise between them, distorting or cropping both in some sense equally. Concretely, two equal area rectangles (with the same center and parallel sides) of different aspect ratios intersect in a rectangle whose aspect ratio is the geometric mean, and their hull (smallest rectangle which contains both of them) likewise has aspect ratio their geometric mean.

In the choice of 16:9 aspect ratio by the SMPTE, balancing 2.35 and 4:3, the geometric mean is $\sqrt{2.35 \times \frac{4}{3}} \approx 1.7701$, and thus $16:9 = 1.77\overline{7}...$ was chosen. This was discovered empirically by Kerns Powers, who cut out rectangles with equal areas and shaped them to match each of the popular aspect ratios. When overlapped with their center points aligned, he found that all of those aspect ratio rectangles fit within an outer rectangle with an aspect ratio of 1.77:1 and all of them also covered a smaller common inner rectangle with the same aspect ratio 1.77:1. The value found by Powers is exactly the geometric mean of the extreme aspect ratios, 4:3 (1.33:1) and CinemaScope (2.35:1), which is coincidentally close to $16:9 (1.77\overline{7}:1)$. Note that the intermediate ratios have no effect on the result, only the two extreme ratios.

Applying the same geometric mean technique to 16:9 and 4:3 approximately yields the 14:9 ($1.55\overline{5}$ ...) aspect ratio, which is likewise used as a compromise between these ratios. In this case 14:9 is exactly the *arithmetic mean* of $16:9$ and $4:3 = 12:9$, since 14 is the average of 16 and 12, while the precise *geometric mean* is $\sqrt{\frac{16}{9} \times \frac{4}{3}} \approx 1.5396 \approx 13.8:9$, but the two different *means*, arithmetic and geometric, are approximately equal because both numbers are sufficiently close to each other (a difference of less than 2%).

Anti-reflective Coatings

In optical coatings, where reflection needs to be minimised between two media of refractive indices n_0 and n_2, the optimum refractive index n_1 of the anti-reflective coating is given by the geometric mean: $n_1 = \sqrt{n_0 n_2}$.

Spectral Flatness

In signal processing, spectral flatness, a measure of how flat or spiky a spectrum is, is defined as the ratio of the geometric mean of the power spectrum to its arithmetic mean.

Geometry

In the case of a right triangle, its altitude is the length of a line extending perpendicularly from the hypotenuse to its 90° vertex. Imagining that this line splits the hypotenuse into two segments, the geometric mean of these segment lengths is the length of the altitude.

In an ellipse, the semi-minor axis is the geometric mean of the maximum and minimum distances of the ellipse from a focus; it is also the geometric mean of the semi-major axis and the semi-latus rectum. The semi-major axis of an ellipse is the geometric mean of the distance from the center to either focus and the distance from the center to either directrix.

Distance to the horizon of a sphere is the geometric mean of the distance to the closest point of the sphere and the distance to the farthest point of the sphere.

Financial

The geometric mean has from time to time been used to calculate financial indices (the averaging is over the components of the index). For example, in the past the FT 30 index used a geometric mean. It is also used in the recently introduced "RPIJ" measure of inflation in the United Kingdom and elsewhere in the European Union.

This has the effect of understating movements in the index compared to using the arithmetic mean.

Harmonic Mean

In mathematics, the harmonic mean (sometimes called the subcontrary mean) is one of several kinds of average, and in particular one of the Pythagorean means. Typically, it is appropriate for situations when the average of rates is desired. The harmonic mean can be expressed as the reciprocal of the arithmetic mean of the reciprocals. As a simple example, the harmonic mean of 1, 2, and 4 is

$$\frac{1}{\frac{1}{3}(\frac{1}{1}+\frac{1}{2}+\frac{1}{4})} = \frac{3}{\frac{1}{1}+\frac{1}{2}+\frac{1}{4}} = \frac{12}{7}.$$

Definition

The harmonic mean H of the positive real numbers $x_1, x_2, \ldots, x_n$ is defined to be

$$H = \frac{n}{\frac{1}{x_1}+\frac{1}{x_2}+\cdots+\frac{1}{x_n}} = \frac{n}{\sum_{i=1}^{n}\frac{1}{x_i}} = \left(\frac{\sum_{i=1}^{n}x_i^{-1}}{n}\right)^{-1}.$$

The third formula in the above equation expresses the harmonic mean as the reciprocal of the arithmetic mean of the reciprocals.

From the following formula:

$$H = \frac{n \cdot \prod\limits_{j=1}^{n} x_j}{\sum\limits_{i=1}^{n} \left\{ \frac{1}{x_i} \prod\limits_{j=1}^{n} x_j \right\}}.$$

it is more apparent that the harmonic mean is related to the arithmetic and geometric means. It is the reciprocal dual of the arithmetic mean for positive inputs:

$$1/H(1/x_1 \ldots 1/x_n) = A(x_1 \ldots x_n)$$

The harmonic mean is a Schur-concave function, and dominated by the minimum of its arguments, in the sense that for any positive set of arguments, $\min(x_1 \ldots x_n) \leq H(x_1 \ldots x_n) \leq n \min(x_1 \ldots x_n)$. Thus, the harmonic mean cannot be made arbitrarily large by changing some values to bigger ones (while having at least one value unchanged).

Relationship with other Means

The harmonic mean is one of the three Pythagorean means. For all *positive* data sets *containing at least one pair of nonequal values*, the harmonic mean is always the least of the three means, while the arithmetic mean is always the greatest of the three and the geometric mean is always in between. (If all values in a nonempty dataset are equal, the three means are always equal to one another; e.g., the harmonic, geometric, and arithmetic means of {2, 2, 2} are all 2.)

It is the special case M_{-1} of the power mean:

$$H(x_1, x_2, \ldots, x_n) = M_{-1}(x_1, x_2, \ldots, x_n) = \frac{n}{x_1^{-1} + x_2^{-1} + \cdots + x_n^{-1}}$$

Since the harmonic mean of a list of numbers tends strongly toward the least elements of the list, it tends (compared to the arithmetic mean) to mitigate the impact of large outliers and aggravate the impact of small ones.

The arithmetic mean is often mistakenly used in places calling for the harmonic mean. In the speed example below for instance, the arithmetic mean of 50 is incorrect, and too big.

The harmonic mean is related to the other Pythagorean means, as seen in the third formula in the above equation. This can be seen by interpreting the denominator to be the arithmetic mean of the product of numbers n times but each time omitting the j-th term. That is, for the first term, we multiply all n numbers except the first; for the second, we multiply all n numbers except the second; and so on. The numerator, excluding the n, which goes with the arithmetic mean, is the geometric mean to the power n. Thus the nth harmonic mean is related to the nth geometric and arithmetic means. The general formula is

$$H(x_1,\ldots,x_n) = \frac{(G(x_1,\ldots,x_n))^n}{A(x_2 x_3 \cdots x_n, x_1 x_3 \cdots x_n, \ldots, x_1 x_2 \cdots x_{n-1})} = \frac{(G(x_1,\ldots,x_n))^n}{A\left(\dfrac{1}{x_1}\displaystyle\prod_{i=1}^{n} x_i, \dfrac{1}{x_2}\displaystyle\prod_{i=1}^{n} x_i, \ldots, \dfrac{1}{x_n}\displaystyle\prod_{i=1}^{n} x_i\right)}.$$

If a set of non-identical numbers is subjected to a mean-preserving spread — that is, two or more elements of the set are "spread apart" from each other while leaving the arithmetic mean unchanged — then the harmonic mean always decreases.

Harmonic Mean of two or three Numbers

Two Numbers

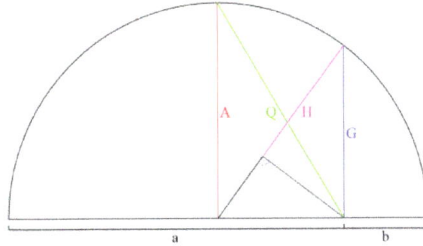

A geometric construction of the three Pythagorean means of two numbers, a and b. The harmonic mean is denoted by H in purple. Q denotes a fourth mean, the quadratic mean. Since a hypotenuse is always longer than a leg of a right triangle, the diagram shows that $Q > A > G > H$.

For the special case of just two numbers, x_1 and x_2, the harmonic mean can be written

$$H = \frac{2 x_1 x_2}{x_1 + x_2}.$$

In this special case, the harmonic mean is related to the arithmetic mean $A = \dfrac{x_1 + x_2}{2}$ and the geometric mean $G = \sqrt{x_1 x_2}$, by

$$H = \frac{G^2}{A} = G \cdot \left(\frac{G}{A}\right).$$

Since $\frac{G}{A} \leq 1$ by the inequality of arithmetic and geometric means, this shows for the $n = 2$ case that $H \leq G$ (a property that in fact holds for all n). It also follows that $G = \sqrt{AH}$, , meaning the two numbers' geometric mean equals the geometric mean of their arithmetic and harmonic means.

Three Numbers

Three positive numbers H, G, and A are respectively the harmonic, geometric, and arithmetic means of three positive numbers if and only if

$$\frac{A^3}{G^3} + \frac{G^3}{H^3} + 1 \le \frac{3}{4}\left(1 + \frac{A}{H}\right)^2.$$

Weighted Harmonic Mean

If a set of weights w_1, ..., w_n is associated to the dataset x_1, ..., x_n, the weighted harmonic mean is defined by

$$H = \frac{\sum_{i=1}^{n} w_i}{\sum_{i=1}^{n} \frac{w_i}{x_i}} = \left(\frac{\sum_{i=1}^{n} w_i x_i^{-1}}{\sum_{i=1}^{n} w_i}\right)^{-1}.$$

The unweighted harmonic mean can be regarded as the special case where all of the weights are equal.

Examples

In physics

In certain situations, especially many situations involving rates and ratios, the harmonic mean provides the truest average. For instance, if a vehicle travels a certain distance at a speed x (e.g., 60 kilometres per hour - km/h) and then the same distance again at a speed y (e.g., 40 km/h), then its average speed is the harmonic mean of x and y (48 km/h), and its total travel time is the same as if it had traveled the whole distance at that average speed. However, if the vehicle travels for a certain amount of *time* at a speed x and then the same amount of time at a speed y, then its average speed is the arithmetic mean of x and y, which in the above example is 50 kilometres per hour. The same principle applies to more than two segments: given a series of sub-trips at different speeds, if each sub-trip covers the same *distance*, then the average speed is the *harmonic* mean of all the sub-trip speeds; and if each sub-trip takes the same amount of *time*, then the average speed is the *arithmetic* mean of all the sub-trip speeds. (If neither is the case, then a weighted harmonic mean or weighted arithmetic mean is needed. For the arithmetic mean, the speed of each portion of the trip is weighted by the duration of that portion, while for the harmonic mean, the corresponding weight is the distance. In both cases, the resulting formula reduces to dividing the total distance by the total time.)

However one may avoid use of the harmonic mean for the case of "weighting by distance". Pose the problem as finding "slowness" of the trip where "slowness" (in hours per kilometre) is the inverse of speed. When trip slowness is found, invert it so as to find the "true" average trip speed. For each trip segment i, the slowness $s_i = 1/\text{speed}_i$. Then take the weighted arithmetic mean of the s_i's weighted by their respective distances (optionally with the weights normalized so they sum to 1 by dividing them by trip length). This gives the true average slowness (in time per kilometre). It turns out that this procedure, which can be done with no knowledge of the harmonic mean, amounts to the same mathematical operations as one would use in solving this problem by using the harmonic mean. Thus it illustrates why the harmonic mean works in this case.

Similarly, if one wishes to estimate the density of an alloy given the densities of its constituent elements and their mass fractions (or, equivalently, percentages by mass), then the predicted density of the alloy (exclusive of typically minor volume changes due to atom packing effects) is the weighted harmonic mean of the individual densities, weighted by mass, rather than the weighted arithmetic mean as one might at first expect. To use the weighted arithmetic mean, the densities would have to be weighted by volume. Applying dimensional analysis to the problem, while labeling the mass units by element and making sure that only like element-masses cancel, makes this clear.

If one connects two electrical resistors in parallel, one having resistance x (e.g., 60 Ω) and one having resistance y (e.g., 40 Ω), then the effect is the same as if one had used two resistors with the same resistance, both equal to the harmonic mean of x and y (48 Ω): the equivalent resistance in either case is 24 Ω (one-half of the harmonic mean). However, if one connects the resistors in series, then the average resistance is the arithmetic mean of x and y (with total resistance equal to the sum of x and y). And, as with the previous example, the same principle applies when more than two resistors are connected, provided that all are in parallel or all are in series.

The same principle applies to capacitors in series.

The "conductivity effective mass" of a semiconductor is also defined as the harmonic mean of the effective masses along the three crystallographic directions.

In other sciences

In computer science, specifically information retrieval and machine learning, the harmonic mean of the precision (true positives per predicted positive) and the recall (true positives per real positive) is often used as an aggregated performance score for the evaluation of algorithms and systems: the F-score (or F-measure). This is used in information retrieval because only the positive class is of relevance and number of negatives is not in general known. It is thus a trade-off as to whether the correct positive predictions should be measured in relation to the number of predicted positives or the number of real positives, so it is measured versus a putative number of positives that is an arithmetic mean of the two possible denominators.

An interesting consequence arises from basic algebra in problems where people or systems work together. As an example, if a gas-powered pump can drain a pool in 4 hours and a battery-powered pump can drain the same pool in 6 hours, then it will take both pumps 6·4/6+4, which is equal to 2.4 hours, to drain the pool together. Interestingly, this is one-half of the harmonic mean of 6 and 4: 2·6·4/6+4 = 4.8. That is the appropriate average for the two types of pump is the harmonic mean, and with one pair of pumps (two pumps) it takes half this harmonic mean time, while with two pairs of pumps (four pumps) it would take a quarter of this harmonic mean time.

In electronics the harmonic mean in the same way gives the average contribution per component for parallel resistance, parallel inductance, serial conductance and serial capacitance.

In hydrology, the harmonic mean is similarly used to average hydraulic conductivity values for flow that is perpendicular to layers (e.g., geologic or soil) - flow parallel to layers uses the arithmetic mean. This apparent difference in averaging is explained by the fact that hydrology uses conductivity, which is the inverse of resistivity.

In sabermetrics, a player's Power–speed number is the harmonic mean of their home run and stolen base totals.

In population genetics, the harmonic mean is used when calculating the effects of fluctuations in generation size on the effective breeding population. This is to take into account the fact that a very small generation is effectively like a bottleneck and means that a very small number of individuals are contributing disproportionately to the gene pool, which can result in higher levels of inbreeding.

In transportation, to find the average speed of a trip over a route divided into constant speed segments (of distance) one must use the weighted harmonic mean (weighted by the distance of each segment). For example, if one travels half-way to a destination at 20 mi/h, and then goes 60 mi/h for the second half of the distance, the average speed is only 30 mi/h (harmonic mean) and not the 40 mi/h (arithmetic mean). This is because it took 3 times as long (in time) to go the first half of the trip distance as it did to go the second half and true average speed is a simple weighted arithmetic mean with the weights being time. Thus 20 mi/h gets 3 times as much weight as 60 mi/h: 3/4·20 + 1/4·60 = 30 mi/h.

When considering fuel economy in automobiles two measures are commonly used – miles per gallon (mpg), and litres per 100 km. As the dimensions of these quantities are the inverse of each other (one is distance per volume, the other volume per distance) when taking the mean value of the fuel-economy of a range of cars one measure will produce the harmonic mean of the other – i.e., converting the mean value of fuel economy expressed in litres per 100 km to miles per gallon will produce the harmonic mean of the fuel economy expressed in miles-per-gallon.

In chemistry and nuclear physics the average mass per particle of a mixture consisting of different species (e.g., molecules or isotopes) is given by the harmonic mean of the individual species' masses weighted by their respective mass fraction.

In Finance

The harmonic mean is the preferable method for averaging multiples, such as the price–earnings ratio, in which price is in the numerator. If these ratios are averaged using an arithmetic mean (a common error), high data points are given greater weights than low data points. The harmonic mean, on the other hand, gives equal weight to each data point. The simple arithmetic mean when applied to non-price normalized ratios such as the P/E is biased upwards and cannot be numerically justified, since it is based on equalized earnings; just as vehicles speeds cannot be averaged for a roundtrip journey.

In Geometry

In any triangle, the radius of the incircle is one-third of the harmonic mean of the altitudes.

For any point P on the minor arc BC of the circumcircle of an equilateral triangle ABC, with distances q and t from B and C respectively, and with the intersection of PA and BC being at a distance y from point P, we have that y is half the harmonic mean of q and t.

In a right triangle with legs a and b and altitude h from the hypotenuse to the right angle, h^2 is half the harmonic mean of a^2 and b^2.

Let t and s ($t > s$) be the sides of the two inscribed squares in a right triangle with hypotenuse c. Then s^2 equals half the harmonic mean of c^2 and t^2.

Let a trapezoid have vertices A, B, C, and D in sequence and have parallel sides AB and CD. Let E be the intersection of the diagonals, and let F be on side DA and G be on side BC such that FEG is parallel to AB and CD. Then FG is the harmonic mean of AB and DC. (This is provable using similar triangles.)

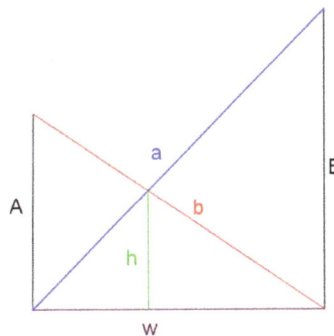

Crossed ladders. h is half the harmonic mean of A and B

In the crossed ladders problem, two ladders lie oppositely across an alley, each with feet at the base of one sidewall, with one leaning against a wall at height *A* and the other leaning against the opposite wall at height *B*, as shown. The ladders cross at a height of *h* above the alley floor. Then *h* is half the harmonic mean of *A* and *B*. This result still holds if the walls are slanted but still parallel and the "heights" *A*, *B*, and *h* are measured as distances from the floor along lines parallel to the walls.

In an ellipse, the semi-latus rectum (the distance from a focus to the ellipse along a line parallel to the minor axis) is the harmonic mean of the maximum and minimum distances of the ellipse from a focus.

Median

$$1, 3, 3, \mathbf{6}, 7, 8, 9$$

$$\text{Median} = \underline{6}$$

$$1, 2, 3, \mathbf{4}, \mathbf{5}, 6, 8, 9$$

$$\text{Median} = (4 + 5) \div 2$$

$$= \underline{4.5}$$

Finding the median in sets of data with an odd and even number of values.

The median is the value separating the higher half of a data sample, a population, or a probability distribution, from the lower half. In simple terms, it may be thought of as the "middle" value of a data set. For example, in the data set {1, 3, 3, 6, 7, 8, 9}, the median is 6, the fourth number in the sample. The median is a commonly used measure of the properties of a data set in statistics and probability theory.

The basic advantage of the median in describing data compared to the mean (often simply described as the "average") is that it is not skewed so much by extremely large or small values, and so it may give a better idea of a 'typical' value. For example, in understanding statistics like household income or assets which vary greatly, a mean may be skewed by a small number of extremely high or low values. Median income, for example, may be a better way to suggest what a 'typical' income is.

Because of this, the median is of central importance in robust statistics, as it is the most resistant statistic, having a breakdown point of 50%: so long as no more than half the data are contaminated, the median will not give an arbitrarily large or small result.

Basic Procedure

The median of a finite list of numbers can be found by arranging all the numbers from smallest to greatest.

If there is an odd number of numbers, the middle one is picked. For example, consider the set of numbers:

$$1, 3, 3, 6, 7, 8, 9$$

This set contains seven numbers. The median is the fourth of them, which is 6.

If there are an even number of observations, then there is no single middle value; the median is then usually defined to be the mean of the two middle values. For example, in the data set:

$$1, 2, 3, 4, 5, 6, 8, 9$$

The median is the mean of the middle two numbers: this is $(4 + 5) \div 2$, which is 4.5. (In more technical terms, this interprets the median as the fully trimmed mid-range.)

The formula used to find the middle number of a data set of n numbers is $(n + 1) \div 2$. This either gives the middle number (for an odd number of values) or the halfway point between the two middle values. For example, with 14 values, the formula will give 7.5, and the median will be taken by averaging the seventh and eighth values.

There is no widely accepted standard notation for the median, but some authors represent the median of a variable x either as $\tilde{x}$ or as $\mu_{1/2}$ sometimes also M. In any of these cases, the use of these or other symbols for the median needs to be explicitly defined when they are introduced.

Discussion

The median is used primarily for skewed distributions, which it summarizes differently from the arithmetic mean. Consider the multiset $\{ 1, 2, 2, 2, 3, 14 \}$. The median is 2 in this case, (as is the mode), and it might be seen as a better indication of central tendency (less susceptible to the exceptionally large value in data) than the arithmetic mean of 4.

Calculation of medians is a popular technique in summary statistics and summarizing statistical data, since it is simple to understand and easy to calculate, while also giving a measure that is more robust in the presence of outlier values than is the mean. The widely cited empirical relationship between the relative locations of the mean and the median for skewed distributions is, however, not generally true. There are, however, various relationships for the *absolute* difference between them.

The median does not identify a specific value within the data set, since more than one value can be at the median level and with an even number of observations no value need be exactly at the value of the median. Nonetheless, the value of the median is uniquely determined with the usual definition. A related concept, in which the outcome is forced to correspond to a member of the sample, is the medoid.

In a population, at most half have values strictly less than the median and at most half have values strictly greater than it. If each group contains less than half the population, then some of the population is exactly equal to the median. For example, if $a < b < c$, then the median of the list $\{a, b, c\}$ is b, and, if $a < b < c < d$, then the median of the list $\{a, b, c, d\}$ is the mean of b and c; i.e., it is $(b + c)/2$. Indeed, as it is based on the middle data in a group, it is not necessary to even know the value of extreme results in order to calculate a median. For example, in a psychology test investigating the time needed to solve a problem, if a small number of people failed to solve the problem at all in the given time a median can still be calculated.

The median can be used as a measure of location when a distribution is skewed, when end-values are not known, or when one requires reduced importance to be attached to outliers, e.g., because they may be measurement errors.

A median is only defined on ordered one-dimensional data, and is independent of any distance metric. A geometric median, on the other hand, is defined in any number of dimensions.

The median is one of a number of ways of summarising the typical values associated with members of a statistical population; thus, it is a possible location parameter. The median is the 2nd quartile, 5th decile, and 50th percentile. Since the median is the same as the *second quartile*, its calculation is illustrated in the article on quartiles. A median can be worked out for ranked but not numerical classes (e.g. working out a median grade when students are graded from A to F), although the result might be halfway between grades if there are an even number of cases.

When the median is used as a location parameter in descriptive statistics, there are several choices for a measure of variability: the range, the interquartile range, the mean absolute deviation, and the median absolute deviation.

For practical purposes, different measures of location and dispersion are often compared on the basis of how well the corresponding population values can be estimated from a sample of data. The median, estimated using the sample median, has good properties in this regard. While it is not usually optimal if a given population distribution is assumed, its properties are always reasonably good. For example, a comparison of the efficiency of candidate estimators shows that the sample mean is more statistically efficient than the sample median when data are uncontaminated by data from heavy-tailed distributions or from mixtures of distributions, but less efficient otherwise, and that the efficiency of the sample median is higher than that for a wide range of distributions. More specifically, the median has a 64% efficiency compared to the minimum-variance mean (for large normal samples), which is to say the variance of the median will be ~50% greater than the variance of the mean

Probability Distributions

For any probability distribution on the real line R with cumulative distribution function F, regardless of whether it is any kind of continuous probability distribution, in particular an absolutely continuous distribution (which has a probability density function), or a discrete probability distribution, a median is by definition any real number m that satisfies the inequalities

$$P(X \leq m) \geq \frac{1}{2} \text{ and } P(X \geq m) \geq \frac{1}{2}$$

or, equivalently, the inequalities

$$\int_{(-\infty, m]} dF(x) \geq \frac{1}{2} \text{ and } \int_{[m, \infty)} dF(x) \geq \frac{1}{2}$$

in which a Lebesgue–Stieltjes integral is used. For an absolutely continuous probability distribution with probability density function f, the median satisfies

$$P(X \leq m) = P(X \geq m) = \int_{-\infty}^{m} f(x)dx = \frac{1}{2}.$$

Any probability distribution on R has at least one median, but there may be more than one median. Where exactly one median exists, statisticians speak of "the median" correctly; even when the median is not unique, some statisticians speak of "the median" informally.

Medians of Particular Distributions

The medians of certain types of distributions can be easily calculated from their parameters; furthermore, they exist even for some distributions lacking a well-defined mean, such as the Cauchy distribution:

- The median of a symmetric distribution which possesses a mean μ also takes the value μ.

  o The median of a normal distribution with mean μ and variance σ^2 is μ. In fact, for a normal distribution, mean = median = mode.

  o The median of a uniform distribution in the interval $[a, b]$ is $(a + b) / 2$, which is also the mean.

- The median of a Cauchy distribution with location parameter x_0 and scale parameter y is x_0, the location parameter.

- The median of a power law distribution x^{-a}, with exponent $a > 1$ is $2^{1/(a-1)} x_{min}$, where x_{min} is the minimum value for which the power law holds

- The median of an exponential distribution with rate parameter λ is the natural logarithm of 2 divided by the rate parameter: $\lambda^{-1} \ln 2$.

- The median of a Weibull distribution with shape parameter k and scale parameter λ is $\lambda (\ln 2)^{1/k}$.

Populations

Optimality Property

The *mean absolute error* of a real variable c with respect to the random variable X is

$$E(|X - c|)$$

Provided that the probability distribution of X is such that the above expectation exists, then m is a median of X if and only if m is a minimizer of the mean absolute error with respect to X. In particular, m is a sample median if and only if m minimizes the arithmetic mean of the absolute deviations.

More generally, a median is defined as a minimum of

$$E(|X - c| - |X|),$$

as discussed below in the section on multivariate medians (specifically, the spatial median).

This optimization-based definition of the median is useful in statistical data-analysis, for example, in k-medians clustering

Unimodal Distributions

It can be shown for a unimodal distribution that the median $\tilde{X}$ and the mean $\bar{X}$ lie within $(3/5)^{1/2} \approx 0.7746$ standard deviations of each other. In symbols,

$$\frac{|\tilde{X} - \bar{X}|}{\sigma} \le (3/5)^{1/2}$$

where $|.|$ is the absolute value.

A similar relation holds between the median and the mode: they lie within $3^{1/2} \approx 1.732$ standard deviations of each other:

$$\frac{|\tilde{X} - \text{mode}|}{\sigma} \le 3^{1/2}.$$

Inequality Relating Means and Medians

If the distribution has finite variance, then the distance between the median and the mean is bounded by one standard deviation.

This bound was proved by Mallows, who used Jensen's inequality twice, as follows. We have

$$|\mu - m| = |\mathrm{E}(X - m)| \leq \mathrm{E}\left(|X - m|\right)$$

$$\leq \mathrm{E}\left(|X - \mu|\right)$$

$$\leq \sqrt{\mathrm{E}((X - \mu)^2)} = \sigma.$$

The first and third inequalities come from Jensen's inequality applied to the absolute-value function and the square function, which are each convex. The second inequality comes from the fact that a median minimizes the absolute deviation function

$$a \mapsto \mathrm{E}(|X - a|).$$

This proof can easily be generalized to obtain a multivariate version of the inequality, as follows:

$$\| \mu - m \| = \| \mathrm{E}(X - m) \| \leq \mathrm{E} \| X - m \|$$
$$\leq \mathrm{E}(\| X - \mu \|)$$
$$\leq \sqrt{\mathrm{E}(\| X - \mu \|^2)} = \sqrt{\mathrm{trace}(\mathrm{var}(X))}$$

where m is a spatial median, that is, a minimizer of the function $a \mapsto \mathrm{E}(\|X - a\|)$. The spatial median is unique when the data-set's dimension is two or more. An alternative proof uses the one-sided Chebyshev inequality; it appears in an inequality on location and scale parameters.

Jensen's Inequality for Medians

Jensen's inequality states that for any random variable x with a finite expectation $E(x)$ and for any convex function f

$$f(E(x)) \leq E(f(x))$$

It has been shown that if x is a real variable with a unique median m and f is a C function then

$$f(m) \leq \text{Median}(f(x))$$

A C function is a real valued function, defined on the set of real numbers R, with the property that for any real t

$$f^{-1}((-\infty, t]) = \{x \in R \mid f(x) \leq t\}$$

is a closed interval, a singleton or an empty set.

Medians for Samples

Efficient Computation of the Sample Median

Even though comparison-sorting n items requires $\Omega(n \log n)$ operations, selection algorithms can compute the k'th-smallest of n items with only $\Theta(n)$ operations. This includes the median, which is the $n/2$'th order statistic (or for an even number of samples, the average of the two middle order statistics).

Selection algorithms still have the downside of requiring $\Omega(n)$ memory, that is, they need to have the full sample (or a linear-sized portion of it) in memory. Because this, as well as the linear time requirement, can be prohibitive, several estimation procedures for the median have been developed. A simple one is the median of three rule, which estimates the median as the median of a three-element subsample; this is commonly used as a subroutine in the quicksort sorting algorithm, which uses an estimate of its input's median. A more robust estimator is Tukey's *ninther*, which is the median of three rule applied with limited recursion: if A is the sample laid out as an array, and

$$\text{med3}(A) = \text{median}(A, A[\frac{n}{2}], A[n]),$$

then

$$\text{ninther}(A) = \text{med3}(\text{med3}(A[1 \ldots \frac{1}{3}n]), \text{med3}(A[\frac{1}{3}n \ldots \frac{2}{3}n]), \text{med3}(A[\frac{2}{3}n \ldots n]))$$

The *remedian* is an estimator for the median that requires linear time but sub-linear memory, operating in a single pass over the sample.

Easy Explanation of the Sample Median

In individual series (if number of observation is very low) first one must arrange all the observations in order. Then count(n) is the total number of observation in given data.

If n is odd then Median (M) = value of $((n + 1)/2)$th item term.

If n is even then Median (M) = value of $[((n)/2)$th item term + $((n)/2 + 1)$th item term $]/2$

For an odd number of values

As an example, we will calculate the sample median for the following set of observations: 1, 5, 2, 8, 7.

Start by sorting the values: 1, 2, 5, 7, 8.

In this case, the median is 5 since it is the middle observation in the ordered list.

The median is the $((n + 1)/2)$th item, where n is the number of values. For example, for the list {1, 2, 5, 7, 8}, we have $n = 5$, so the median is the $((5 + 1)/2)$th item.

median = (6/2)th item

median = 3rd item

median = 5

For an even number of values

As an example, we will calculate the sample median for the following set of observations: 1, 6, 2, 8, 7, 2.

Start by sorting the values: 1, 2, 2, 6, 7, 8.

In this case, the arithmetic mean of the two middlemost terms is $(2 + 6)/2 = 4$. Therefore, the median is 4 since it is the arithmetic mean of the middle observations in the ordered list.

We also use this formula MEDIAN = {$(n + 1)/2$}th item . n = number of values

As above example 1, 2, 2, 6, 7, 8 $n = 6$ Median = {$(6 + 1)/2$}th item = 3.5th item. In this case, the median is average of the 3rd number and the next one (the fourth number). The median is $(2 + 6)/2$ which is 4.

Sampling Distribution

The distribution of both the sample mean and the sample median were determined by Laplace. The distribution of the sample median from a population with a density function $f(x)$ is asymptotically normal with mean m and variance

$$\frac{1}{4nf(m)^2}$$

where m is the median value of distribution and n is the sample size. In practice this may be difficult to estimate as the density function is usually unknown.

These results have also been extended. It is now known for the p-th quantile that the distribution of the sample p-th quantile is asymptotically normal around the p-th quantile with variance equal to

$$\frac{p(1-p)}{nf(x_p)^2}$$

where $f(x_p)$ is the value of the distribution density at the p-th quantile.

In the case of a discrete variable, the sampling distribution of the median for small-samples can be investigated as follows. We take the sample size to be an odd number N = 2 n + 1 If a given value v is to be the median of the sample then two conditions must be satisfied. The first is that at most n observations can have a value of v – 1or less. The second is that at most n observations can have a value of v + 1 or more. Let i be the number of observations which have a value of v – 1 or less and let k be the number of observations which have a value of v + 1or more. Then i and k both have a minimum value of 0 and a maximum of n . If an observation has a value below v , it is not relevant how far below v it is and conversely, if an observation has a value above v , it is not relevant how far above v it is. We can therefore represent the observations as following a trinomial distribution with probabilities F (v – 1) , f (v) and 1 – F (v) . The probability that the median m will have a value vis then given by

$$\Pr(m=v) = \sum_{i=0}^{n}\sum_{k=0}^{n} \frac{N!}{i!(N-i-k)!k!}[F(v-1)]^i[f(v)]^{N-i-k}[1-F(v)]^k.$$

Summing this over all values of vdefines a proper distribution and gives a unit sum. In practice, the function f (v) will often not be known but it can be estimated from an observed frequency distribution. An example is given in the following table where the actual distribution is not known but a sample of 3,800 observations allows a sufficiently accurate assessment of f (v).

v	0	0.5	1	1.5	2	2.5	3	3.5	4	4.5	5
f(v)	0.000	0.008	0.010	0.013	0.083	0.108	0.328	0.220	0.202	0.023	0.005
F(v)	0.000	0.008	0.018	0.031	0.114	0.222	0.550	0.770	0.972	0.995	1.000

Using these data it is possible to investigate the effect of sample size on the standard errors of the mean and median. The observed mean is 3.16, the observed raw median is 3 and the observed interpolated median is 3.174. The following table gives some comparison statistics. The standard error of the median is given both from the above expression for p r (m = v) and from the asymptotic approximation given earlier.

Sample size Statistic	3	9	15	21
Expected value of median	3.198	3.191	3.174	3.161

Standard error of median (above formula)	0.482	0.305	0.257	0.239
Standard error of median (asymptotic approximation)	0.879	0.508	0.393	0.332
Standard error of mean	0.421	0.243	0.188	0.159

The expected value of the median falls slightly as sample size increases while, as would be expected, the standard errors of both the median and the mean are proportionate to the inverse square root of the sample size. The asymptotic approximation errs on the side of caution by overestimating the standard error.

In the case of a continuous variable, the following argument can be used. If a given value v is to be the median, then one observation must take the value v. The elemental probability of this is f (v) d v. Then, of the remaining 2 n observations, exactly n of them must be above v and the remaining n below. The probability of this is the n th term of a binomial distribution with parameters F (v) and 2 n. Finally we multiply by 2 n + 1 since any of the observations in the sample can be the median observation. Hence the elemental probability of the median at the point v is given by

$$f(v)\frac{(2n)!}{n!n!}[F(v)]^n[1-F(v)]^n(2n+1)dv.$$

Now we introduce the beta function. For integer arguments α and β, this can be expressed as $B(\alpha,\beta)=(\alpha-1)!(\beta-1)!/(\alpha+\beta-1)!$. Also, we note that f (v) = d F (v) / d v. Using these relationships and setting both α and β equal to (n + 1) allows the last expression to be written as

$$\frac{[F(v)]^n[1-F(v)]^n}{B(n+1,n+1)}dF(v)$$

Hence the density function of the median is a symmetric beta distribution over the unit interval which supports F (v) . Its mean, as we would expect, is 0.5 and its standard deviation (which is the standard error of the sample median) is 1 / (2 N + 2). However this finding can only be used if (i) f (v) is known or can be assumed (ii) f (v) can be integrated to find F (v) and (iii) F (v) can in turn be inverted. This will not always be the case and even when it is, the cut-off points for f (v) can be calculated directly without recourse to the distribution of the median on the unit interval. So while interesting in theory, this result is not much use in practice.

Estimation of Variance from Sample Data

The value of $(2f(x))^{-2}$ —the asymptotic value of $n^{-\frac{1}{2}}(\nu-m)$ where ν is the population median—has been studied by several authors. The standard 'delete one' jackknife method produces inconsistent results. An alternative—the 'delete k' method—where k

grows with the sample size has been shown to be asymptotically consistent. This method may be computationally expensive for large data sets. A bootstrap estimate is known to be consistent, but converges very slowly (order of $n^{-\frac{1}{4}}$)). Other methods have been proposed but their behavior may differ between large and small samples.

Efficiency

The efficiency of the sample median, measured as the ratio of the variance of the mean to the variance of the median, depends on the sample size and on the underlying population distribution. For a sample of size $N = 2n + 1$ from the normal distribution, the ratio is

$$\frac{4n}{\pi(2n+1)}$$

For large samples (as n tends to infinity) this ratio tends to $\frac{2}{\pi}$.

Other Estimators

For univariate distributions that are *symmetric* about one median, the Hodges–Lehmann estimator is a robust and highly efficient estimator of the population median.

If data are represented by a statistical model specifying a particular family of probability distributions, then estimates of the median can be obtained by fitting that family of probability distributions to the data and calculating the theoretical median of the fitted distribution. Pareto interpolation is an application of this when the population is assumed to have a Pareto distribution.

Coefficient of Dispersion

The coefficient of dispersion (CD) is defined as the ratio of the average absolute deviation from the median to the median of the data. It is a statistical measure used by the states of Iowa, New York and South Dakota in estimating dues taxes. In symbols

$$CD = \frac{1}{n} \frac{\sum |m - x|}{m}$$

where n is the sample size, m is the sample median and x is a variate. The sum is taken over the whole sample.

Confidence intervals for a two sample test where the sample sizes are large have been derived by Bonett and Seier This test assumes that both samples have the same median but differ in the dispersion around it. The confidence interval (CI) is bounded inferiorly by

$$\exp\left[\log\left(\frac{t_a}{t_b}\right) - z_\alpha\left(\text{var}\left[\log\left(\frac{t_a}{t_b}\right)\right]\right)^{0.5}\right]$$

where t_j is the mean absolute deviation of the jth sample, var() is the variance and z_a is the value from the normal distribution for the chosen value of a: for $a = 0.05$, $z_a = 1.96$. The following formulae are used in the derivation of these confidence intervals

$$\text{var}[\log(t_a)] = \frac{\left(\frac{s_a^2}{t_a^2} + \left(\frac{x_a - \overline{x}}{t_a}\right)^2 - 1\right)}{n}$$

$$\text{var}\left[\log\left(\frac{t_a}{t_b}\right)\right] = \text{var}[\log(t_a)] + \text{var}[\log(t_b)] - 2r(\text{var}[\log(t_a)]\,\text{var}[\log(t_b)])^{0.5}$$

where r is the Pearson correlation coefficient between the squared deviation scores

$$d_{ia} = |x_{ia} - \overline{x}_a| \text{ and } d_{ib} = |x_{ib} - \overline{x}_b|$$

a and b here are constants equal to 1 and 2, x is a variate and s is the standard deviation of the sample.

Multivariate Median

Previously, this article discussed the concept of a univariate median for a one-dimensional object (population, sample). When the dimension is two or higher, there are multiple concepts that extend the definition of the univariate median; each such multivariate median agrees with the univariate median when the dimension is exactly one. In higher dimensions, however, there are several multivariate medians.

Marginal Median

The marginal median is defined for vectors defined with respect to a fixed set of coordinates. A marginal median is defined to be the vector whose components are univariate medians. The marginal median is easy to compute, and its properties were studied by Puri and Sen.

Spatial Median (L1 median)

In a normed vector space of dimension two or greater, the "spatial median" minimizes the expected distance

$$a \mapsto 1/N \sum_n (\|x_n - a\|),$$

where x_n and a are vectors. The spatial median is unique when the data-set's dimension is two or more and the norm is the Euclidean norm (or another strictly convex norm). The spatial median is also called the L1 median, even when the norm is Euclidean. Other names are used especially for finite sets of points: geometric median, Fermat point (in mechanics), or Weber or Fermat-Weber point (in geographical location theory).

More generally, a spatial median is defined as a minimizer of

$$a \mapsto 1/N \sum_n (\|x_n - a\| - \|x_n\|) \; ;$$

this general definition is convenient for defining a spatial median of a population in a finite-dimensional normed space, for example, for distributions without a finite mean. Spatial medians are defined for random vectors with values in a Banach space.

The spatial median is a robust and highly efficient estimator of a central tendency of a population.

Other Multivariate Medians

An alternative generalization of the spatial median in higher dimensions that does not relate to a particular metric is the centerpoint.

Other Median-related Concepts

Pseudo-median

For univariate distributions that are *symmetric* about one median, the Hodges–Lehmann estimator is a robust and highly efficient estimator of the population median; for non-symmetric distributions, the Hodges–Lehmann estimator is a robust and highly efficient estimator of the population *pseudo-median*, which is the median of a symmetrized distribution and which is close to the population median. The Hodges–Lehmann estimator has been generalized to multivariate distributions.

Variants of Regression

The Theil–Sen estimator is a method for robust linear regression based on finding medians of slopes.

Median Filter

In the context of image processing of monochrome raster images there is a type of noise, known as the salt and pepper noise, when each pixel independently becomes black (with some small probability) or white (with some small probability), and is unchanged otherwise (with the probability close to 1). An image constructed of median values of neighborhoods (like 3×3 square) can effectively reduce noise in this case.

Cluster Analysis

In cluster analysis, the k-medians clustering algorithm provides a way of defining clusters, in which the criterion of maximising the distance between cluster-means that is used in k-means clustering, is replaced by maximising the distance between cluster-medians.

Median–median Line

This is a method of robust regression. The idea dates back to Wald in 1940 who suggested dividing a set of bivariate data into two halves depending on the value of the independent parameter x: a left half with values less than the median and a right half with values greater than the median. He suggested taking the means of the dependent y and independent x variables of the left and the right halves and estimating the slope of the line joining these two points. The line could then be adjusted to fit the majority of the points in the data set.

Nair and Shrivastava in 1942 suggested a similar idea but instead advocated dividing the sample into three equal parts before calculating the means of the subsamples. Brown and Mood in 1951 proposed the idea of using the medians of two subsamples rather the means. Tukey combined these ideas and recommended dividing the sample into three equal size subsamples and estimating the line based on the medians of the subsamples.

Median-unbiased Estimators

Any *mean*-unbiased estimator minimizes the risk (expected loss) with respect to the squared-error loss function, as observed by Gauss. A *median*-unbiased estimator minimizes the risk with respect to the absolute-deviation loss function, as observed by Laplace. Other loss functions are used in statistical theory, particularly in robust statistics.

The theory of median-unbiased estimators was revived by George W. Brown in 1947:

An estimate of a one-dimensional parameter θ will be said to be median-unbiased if, for fixed θ, the median of the distribution of the estimate is at the value θ; i.e., the estimate underestimates just as often as it overestimates. This requirement seems for most purposes to accomplish as much as the mean-unbiased requirement and has the additional property that it is invariant under one-to-one transformation.

Further properties of median-unbiased estimators have been reported. In particular, median-unbiased estimators exist in cases where mean-unbiased and maximum-likelihood estimators do not exist. Median-unbiased estimators are invariant under one-to-one transformations.

History

The idea of the median appeared in the 13th century in the Talmud (further for possible older mentions)

The idea of the median also appeared later in Edward Wright's book on navigation (*Certaine Errors in Navigation*) in 1599 in a section concerning the determination of location with a compass. Wright felt that this value was the most likely to be the correct value in a series of observations.

In 1757, Roger Joseph Boscovich developed a regression method based on the L1 norm and therefore implicitly on the median.

In 1774, Laplace suggested the median be used as the standard estimator of the value of a posterior pdf. The specific criteria was to minimize the expected magnitude of the error; $|\alpha - \alpha^*|$ where α^* is the estimate and α is the true value. Laplaces's criterion was generally rejected for 150 years in favor of the least squares method of Gauss and Legendre which minimizes $(\alpha - \alpha^*)^2$ to obtain the mean. The distribution of both the sample mean and the sample median were determined by Laplace in the early 1800s.

Antoine Augustin Cournot in 1843 was the first to use the term *median (valeur médiane)* for the value that divides a probability distribution into two equal halves. Gustav Theodor Fechner used the median (*Centralwerth*) in sociological and psychological phenomena. It had earlier been used only in astronomy and related fields. Gustav Fechner popularized the median into the formal analysis of data, although it had been used previously by Laplace.

Francis Galton used the English term *median* in 1881, having earlier used the terms *middle-most value* in 1869 and the *medium* in 1880.

Mode (Statistics)

The mode is the value that appears most often in a set of data. The mode of a discrete probability distribution is the value x at which its probability mass function takes its maximum value. In other words, it is the value that is most likely to be sampled. The mode of a continuous probability distribution is the value x at which its probability density function has its maximum value, so the mode is at the peak.

Like the statistical mean and median, the mode is a way of expressing, in a single number, important information about a random variable or a population. The numerical value of the mode is the same as that of the mean and median in a normal distribution, and it may be very different in highly skewed distributions.

The mode is not necessarily unique to a given distribution, since the probability mass function or probability density function may take the same maximum value at several points x_1, x_2, etc. The most extreme case occurs in uniform distributions, where all values occur equally frequently. When a probability density function has multiple local maxima it is common to refer to all of the local maxima as modes of the distribution. Such a continuous distribution is called multimodal (as opposed to unimodal).

In symmetric unimodal distributions, such as the normal distribution, the mean (if defined), median and mode all coincide. For samples, if it is known that they are drawn from a symmetric distribution, the sample mean can be used as an estimate of the population mode.

Mode of a Sample

The mode of a sample is the element that occurs most often in the collection. For example, the mode of the sample [1, 3, 6, 6, 6, 6, 7, 7, 12, 12, 17] is 6. Given the list of data [1, 1, 2, 4, 4] the mode is not unique - the dataset may be said to be bimodal, while a set with more than two modes may be described as multimodal.

For a sample from a continuous distribution, such as [0.935..., 1.211..., 2.430..., 3.668..., 3.874...], the concept is unusable in its raw form, since no two values will be exactly the same, so each value will occur precisely once. In order to estimate the mode, the usual practice is to discretize the data by assigning frequency values to intervals of equal distance, as for making a histogram, effectively replacing the values by the midpoints of the intervals they are assigned to. The mode is then the value where the histogram reaches its peak. For small or middle-sized samples the outcome of this procedure is sensitive to the choice of interval width if chosen too narrow or too wide; typically one should have a sizable fraction of the data concentrated in a relatively small number of intervals (5 to 10), while the fraction of the data falling outside these intervals is also sizable. An alternate approach is kernel density estimation, which essentially blurs point samples to produce a continuous estimate of the probability density function which can provide an estimate of the mode.

The following MATLAB (or Octave) code example computes the mode of a sample:

```
X = sort(x);

indices  = find(diff([X; realmax]) > 0); % indices where repeated values change

[modeL,i] =  max (diff([0; indices]));    % longest persistence length of repeated values

mode     = X(indices(i));
```

The algorithm requires as a first step to sort the sample in ascending order. It then computes the discrete derivative of the sorted list, and finds the indices where this derivative is positive. Next it computes the discrete derivative of this set of indices, locating the maximum of this derivative of indices, and finally evaluates the sorted sample at the point where that maximum occurs, which corresponds to the last member of the stretch of repeated values.

Comparison of Mean, Median and Mode

Comparison of common averages of values { 1, 2, 2, 3, 4, 7, 9 }			
Type	**Description**	**Example**	**Result**
Arithmetic mean	Sum of values of a data set divided by number of values: $$\bar{x} = \frac{1}{n} \sum_{i=1}^{n} x_i$$	(1+2+2+3+4+7+9) / 7	4
Median	Middle value separating the greater and lesser halves of a data set	1, 2, 2, 3, 4, 7, 9	3
Mode	Most frequent value in a data set	1, 2, 2, 3, 4, 7, 9	2

Use

Unlike mean and median, the concept of mode also makes sense for "nominal data" (i.e., not consisting of numerical values in the case of mean, or even of ordered values in the case of median). For example, taking a sample of Korean family names, one might find that "Kim" occurs more often than any other name. Then "Kim" would be the mode of the sample. In any voting system where a plurality determines victory, a single modal value determines the victor, while a multi-modal outcome would require some tie-breaking procedure to take place.

Unlike median, the concept of mode makes sense for any random variable assuming values from a vector space, including the real numbers (a one-dimensional vector space) and the integers (which can be considered embedded in the reals). For example, a distribution of points in the plane will typically have a mean and a mode, but the concept of median does not apply. The median makes sense when there is a linear order on the possible values. Generalizations of the concept of median to higher-dimensional spaces are the geometric median and the centerpoint.

Uniqueness and Definedness

For some probability distributions, the expected value may be infinite or undefined, but if defined, it is unique. The mean of a (finite) sample is always defined. The median is the value such that the fractions not exceeding it and not falling below it are each at least 1/2. It is not necessarily unique, but never infinite or totally undefined. For a data sample it is the "halfway" value when the list of values is ordered in increasing value, where usually for a list of even length the numerical average is taken of the two values closest to "halfway". Finally, as said before, the mode is not necessarily unique. Certain pathological distributions (for example, the Cantor distribution) have no defined mode at all. For a finite data sample, the mode is one (or more) of the values in the sample.

Properties

Assuming definedness, and for simplicity uniqueness, the following are some of the most interesting properties.

- All three measures have the following property: If the random variable (or each value from the sample) is subjected to the linear or affine transformation which replaces X by $aX+b$, so are the mean, median and mode.

- However, if there is an arbitrary monotonic transformation, only the median follows; for example, if X is replaced by $\exp(X)$, the median changes from m to $\exp(m)$ but the mean and mode won't.

- Except for extremely small samples, the mode is insensitive to "outliers" (such as occasional, rare, false experimental readings). The median is also very robust in the presence of outliers, while the mean is rather sensitive.

- In continuous unimodal distributions the median lies, as a rule of thumb, between the mean and the mode, about one third of the way going from mean to mode. In a formula, median $\approx (2 \times \text{mean} + \text{mode})/3$. This rule, due to Karl Pearson, often applies to slightly non-symmetric distributions that resemble a normal distribution, but it is not always true and in general the three statistics can appear in any order.

- For unimodal distributions, the mode is within standard deviations of the mean, and the root mean square deviation about the mode is between the standard deviation and twice the standard deviation.

Example for a Skewed Distribution

An example of a skewed distribution is personal wealth: Few people are very rich, but among those some are extremely rich. However, many are rather poor.

A well-known class of distributions that can be arbitrarily skewed is given by the log-normal distribution. It is obtained by transforming a random variable X having a normal distribution into random variable $Y = e^X$. Then the logarithm of random variable Y is normally distributed, hence the name.

Taking the mean μ of X to be 0, the median of Y will be 1, independent of the standard deviation σ of X. This is so because X has a symmetric distribution, so its median is also 0. The transformation from X to Y is monotonic, and so we find the median $e^0 = 1$ for Y.

When X has standard deviation $\sigma = 0.25$, the distribution of Y is weakly skewed. Using formulas for the log-normal distribution, we find:

$$
\begin{aligned}
\text{mean} &= e^{\mu+\sigma^2/2} &= e^{0+0.25^2/2} &\approx 1.032 \\
\text{mode} &= e^{\mu-\sigma^2} &= e^{0-0.25^2} &\approx 0.939 \\
\text{median} &= e^{\mu} &= e^{0} &= 1
\end{aligned}
$$

Indeed, the median is about one third on the way from mean to mode.

When X has a larger standard deviation, $\sigma = 1$, the distribution of Y is strongly skewed. Now

$$
\begin{aligned}
\text{mean} &= e^{\mu+\sigma^2/2} &= e^{0+1^2/2} &\approx 1.649 \\
\text{mode} &= e^{\mu-\sigma^2} &= e^{0-1^2} &\approx 0.368 \\
\text{median} &= e^{\mu} &= e^{0} &= 1
\end{aligned}
$$

Here, Pearson's rule of thumb fails.

Van Zwet Condition

Van Zwet derived an inequality which provides sufficient conditions for this inequality to hold. The inequality

Mode ≤ Median ≤ Mean

holds if

F(Median - x) + F(Median + x) ≥ 1

for all x where F() is the cumulative distribution function of the distribution.

Unimodal Distributions

It can be shown for a unimodal distribution that the median $\tilde{X}$ and the mean $\bar{X}$ lie within $(3/5)^{1/2} \approx 0.7746$ standard deviations of each other. In symbols,

$$
\frac{\left|\tilde{X} - \bar{X}\right|}{\sigma} \le (3/5)^{1/2}
$$

where $|\cdot|$ is the absolute value.

A similar relation holds between the median and the mode: they lie within $3^{1/2} \approx 1.732$ standard deviations of each other:

$$
\frac{\left|\tilde{X} - \text{mode}\right|}{\sigma} \le 3^{1/2}.
$$

Confidence Interval for the Mode with a Single Data Point

It is a common but false belief that from a single observation x we can not gain information about the variability in the population and that consequently that finite length confidence intervals for mean and/or variance are impossible even in principle.

It is possible for an unknown unimodal distribution to estimate a confidence interval for the mode with a sample size of 1. This was first shown by Abbot and Rosenblatt and extended by Blachman and Machol. This confidence interval can be sharpened if the distribution can be assumed to be symmetrical. It is further possible to sharpen this interval if the distribution is normally distributed.

Let the confidence interval be 1 - α. Then the confidence intervals for the general, symmetric and normally distributed variates respectively are

$$X \pm \left(\frac{2}{\alpha} - 1\right)|X - \theta|$$

$$X \pm \left(\frac{1}{\alpha} - 1\right)|X - \theta|$$

$$X \pm \left(\frac{0.484}{\alpha} - 1\right)|X - \theta|$$

where X is the variate, θ is the mode and $|\cdot|$ is the absolute value.

These estimates are conservative. The confidence intervals for the mode at the 90% level given by these estimators are $X \pm 19|X - \theta|$, $X \pm 9|X - \theta|$ and $X \pm 5.84|X - \theta|$ for the general, symmetric and normally distributed variates respectively. The 95% confidence interval for a normally distributed variate is given by $X \pm 10.7|X - \theta|$. It may be worth noting that the mean and the mode coincide if the variates are normally distributed.

The 95% bound for a normally distributed variate has been improved and is now known to be $X \pm 9.68|X - \theta|$ The bound for a 99% confidence interval is $X \pm 48.39|X - \theta\rangle|$

Note

Machol has shown that given a known density symmetrical about 0 that given a single sample value (x) that the 90% confidence intervals of population mean are

$$x \pm 5|x - v|$$

where v is the population median.

If the precise form of the distribution is not known but it is known to be symmetrical about zero then we have

$$P(X-k\,|\,X-a\,|\le \mu \le X+k\,|\,X-a\,|)\ge 1-\frac{1}{1+k}$$

where X is the variate, μ is the population mean and a and k are arbitrary real numbers.

It is also possible to estimate a confidence interval for the standard deviation from a single observation if the distribution is symmetrical about o. For a normal distribution the with an unknown variance and a single data point (X) the 90%, 95% and 99% confidence intervals for the standard deviation are [o, 8|X|], [o, 17|X|] and [o, 70|X|]. These intervals may be shorted if the mean is known to be bounded by a multiple of the standard deviation.

If the distribution is known to be normal then it is possible to estimate a confidence interval for the mean and variance from a simple value. The 90% confidence intervals are

$$X-23.3\,|\,X\,|\le \mu \le X+23.3\,|\,X\,|$$

$$\sigma \le 10\,|\,X\,|$$

The confidence intervals can be estimated for any chosen range.

This method is not limited to the normal distribution but can be used with any known distribution.

Statistical Tests

These estimators have been used to create hypothesis tests for simple samples from normal or symmetrical unimodal distributions. Let the distribution have an assumed mean (μ_0). The null hypothesis is that the assumed mean of the distribution lies within the confidence interval of the sample mean (m). The null hypothesis is accepted if

$$\mu_0 < \frac{x+m}{2} \pm k\,|\,x-m\,|$$

where x is the value of the sample and k is a constant. The null hypothesis is rejected if

$$\mu_0 > \frac{x+m}{2} \pm k\,|\,x-m\,|$$

The value of k depends on the choice of confidence interval and the nature of the assumed distribution.

If the distribution is assumed or is known to be normal then the values of k for the 50%, 66.6%, 75%, 80%, 90%, 95% and 99% confidence intervals are 0.50, 1.26, 1.80, 2.31, 4.79, 9.66 and 48.39 respectively.

If the distribution is assumed or known to be unimodal and symmetrical but not normal then the values of k for the 50%, 66.6%, 75%, 80%, 90%, 95% and 99% confidence intervals are 0.50, 1.87, 2.91, 3.94, 8.97, 18.99, 99.00 respectively.

To see how this test works we assume or know *a priori* that the population from which the sample is drawn has a mean of μ_0 and that the population has a symmetrical unimodal distribution - a class that includes the normal distribution. We wish to know if the mean estimated from the sample is representative of the population at a pre chosen level of confidence.

Assume that the distribution is normal and let the confidence interval be 95%. Then k = 9.66.

Assuming that the sample is representative of the population, the sample mean (m) will then lie within the range determined from the formula:

$$\mu_0 < \frac{x+m}{2} \pm 9.66 \, | \, x - m \, |$$

If subsequent sampling shows that the sample mean lies outside these parameters the sample mean is to be considered to differ significantly from the population mean.

History

The term mode originates with Karl Pearson in 1895.

References

- Jacobs, Harold R. (1994). Mathematics: A Human Endeavor (Third ed.). W. H. Freeman. p. 547. ISBN 0-7167-2426-X.

- Foerster, Paul A. (2006). Algebra and Trigonometry: Functions and Applications, Teacher's Edition (Classics ed.). Upper Saddle River, NJ: Prentice Hall. p. 573. ISBN 0-13-165711-9.

- Medhi, Jyotiprasad (1992). Statistical Methods: An Introductory Text. New Age International. pp. 53–58. ISBN 9788122404197.

- Feller, William (1950). Introduction to Probability Theory and its Applications, Vol I. Wiley. p. 221. ISBN 0471257087.

- "Fairness Opinions: Common Errors and Omissions". The Handbook of Business Valuation and Intellectual Property Analysis. McGraw Hill. 2004. ISBN 0-07-142967-0.

- Posamentier, Alfred S.; Salkind, Charles T. (1996). Challenging Problems in Geometry (Second ed.). Dover. p. 172. ISBN 0-486-69154-3.

- David J. Sheskin (27 August 2003). Handbook of Parametric and Nonparametric Statistical Procedures: Third Edition. CRC Press. pp. 7–. ISBN 978-1-4200-3626-8. Retrieved 25 February 2013.

- Derek Bissell (1994). Statistical Methods for Spc and Tqm. CRC Press. pp. 26–. ISBN 978-0-412-39440-9. Retrieved 25 February 2013.

Measures of Statistical Deviation: An Overview

In statistics, deviation is a measure of difference between the value observed and some other value. Some of the topics discussed in this chapter include variance, standard deviation, interquartile range, statistical range and mean absolute difference. The topics discussed in the chapter are of great importance to broaden the existing knowledge on statistical deviation.

Variance

In probability theory and statistics, variance is the expectation of the squared deviation of a random variable from its mean, and it informally measures how far a set of (random) numbers are spread out from their mean. The variance has a central role in statistics. It is used in descriptive statistics, statistical inference, hypothesis testing, goodness of fit, Monte Carlo sampling, amongst many others. This makes it a central quantity in numerous fields such as physics, biology, chemistry, economics, and finance. The variance is the square of the standard deviation, the second central moment of a distribution, and the covariance of the random variable with itself, and it is often represented by σ^2 s² or $\text{Var}(X)$.

Definition

The variance of a random variable X is the expected value of the squared deviation from the mean of X, $\mu = \text{E}[X]$:

$$\text{Var}(X) = \text{E}\left[(X - \mu)^2\right].$$

This definition encompasses random variables that are generated by processes that are discrete, continuous, neither, or mixed. The variance can also be thought of as the covariance of a random variable with itself:

$$\text{Var}(X) = \text{Cov}(X, X).$$

The variance is also equivalent to the second cumulant of a probability distribution that generates X. The variance is typically designated as $\text{Var}(X)$, σ_X^2, or simply σ^2 (pronounced "sigma squared"). The expression for the variance can be expanded:

$$\text{Var}(X) = \text{E}\left[(X - \text{E}[X])^2\right]$$

$$= \text{E}\left[X^2 - 2X\,\text{E}[X] + (\text{E}[X])^2\right]$$

$$= \text{E}\left[X^2\right] - 2\,\text{E}[X]\,\text{E}[X] + (\text{E}[X])^2$$

$$= \text{E}\left[X^2\right] - (\text{E}[X])^2$$

A mnemonic for the above expression is "mean of square minus square of mean". On computational floating point arithmetic, this equation should not be used, because it suffers from catastrophic cancellation if the two components of the equation are similar in magnitude. There exist numerically stable alternatives.

Continuous Random Variable

If the random variable X represents samples generated by a continuous distribution with probability density function $f(x)$, then the population variance is given by

$$\text{Var}(X) = \sigma^2 = \int (x - \mu)^2 f(x)dx = \int x^2 f(x)dx - 2\mu\int x f(x)dx + \int \mu^2 f(x)dx = \int x^2 f(x)dx - \mu^2$$

where μ is the expected value,

$$\mu = \int x f(x)dx$$

and where the integrals are definite integrals taken for x ranging over the range of X.

If a continuous distribution does not have an expected value, as is the case for the Cauchy distribution, it does not have a variance either. Many other distributions for which the expected value does exist also do not have a finite variance because the integral in the variance definition diverges. An example is a Pareto distribution whose index k satisfies $1 < k \leq 2$.

Discrete Random Variable

If the generator of random variable X is discrete with probability mass function $x_1 \mapsto p_1, x_2 \mapsto p_2, \ldots, x_n \mapsto p_n$ then

$$\text{Var}(X) = \sum_{i=1}^{n} p_i \cdot (x_i - \mu)^2,$$

or equivalently

$$\text{Var}(X) = \sum_{i=1}^{n} p_i x_i^2 - \mu^2,$$

where μ is the expected value, i.e.

$$\mu = \sum_{i=1}^{n} p_i \cdot x_i.$$

(When such a discrete weighted variance is specified by weights whose sum is not 1, then one divides by the sum of the weights.)

The variance of a set of n equally likely values can be written as

$$\text{Var}(X) = \frac{1}{n} \sum_{i=1}^{n} (x_i - \mu)^2.$$

where μ is the expected value, i.e.,

$$\mu = \frac{1}{n} \sum_{i=1}^{n} x_i$$

The variance of a set of n equally likely values can be equivalently expressed, without directly referring to the mean, in terms of squared deviations of all points from each other:

$$\text{Var}(X) = \frac{1}{n^2} \sum_{i=1}^{n} \sum_{j=1}^{n} \frac{1}{2} (x_i - x_j)^2 = \frac{1}{n^2} \sum_{i} \sum_{j>i} (x_i - x_j)^2.$$

Examples

Normal Distribution

The normal distribution with parameters μ and σ is a continuous distribution whose probability density function is given by

$$f(x) = \frac{1}{\sqrt{2\pi\sigma^2}} e^{-\frac{(x-\mu)^2}{2\sigma^2}}.$$

In this distribution, E(X) = μ and the variance Var(X) is related with σ via

$$\text{Var}(X) = \int_{-\infty}^{\infty} \frac{(x-\mu)^2}{\sqrt{2\pi\sigma^2}} e^{-\frac{(x-\mu)^2}{2\sigma^2}} dx = \sigma^2.$$

The role of the normal distribution in the central limit theorem is in part responsible for the prevalence of the variance in probability and statistics.

Exponential Distribution

The exponential distribution with parameter λ is a continuous distribution whose support is the semi-infinite interval $[0, \infty[$. Its probability density function is given by

$$f(x) = \lambda e^{-\lambda x},$$

and it has expected value $\mu = \lambda^{-1}$. The variance is equal to

$$\text{Var}(X) = \int_0^\infty (x - \lambda^{-1})^2 \lambda e^{-\lambda x} dx = \lambda^{-2}.$$

So for an exponentially distributed random variable, $\sigma^2 = \mu^2$.

Poisson Distribution

The Poisson distribution with parameter λ is a discrete distribution for $k = 0, 1, 2, \ldots$. Its probability mass function is given by

$$p(k) = \frac{\lambda^k}{k!} e^{-\lambda},$$

and it has expected value $\mu = \lambda$. The variance is equal to

$$\text{Var}(X) = \sum_{k=0}^\infty \frac{\lambda^k}{k!} e^{-\lambda} (k - \lambda)^2 = \lambda,$$

So for a Poisson-distributed random variable, $\sigma^2 = \mu$.

Binomial Distribution

The binomial distribution with parameters n and p is a discrete distribution for $k = 0, 1, 2, \ldots, n$. Its probability mass function is given by

$$p(k) = \binom{n}{k} p^k (1 - p)^{n-k},$$

and it has expected value $\mu = np$. The variance is equal to

$$\text{Var}(X) = \sum_{k=0}^n \binom{n}{k} p^k (1 - p)^{n-k} (k - np)^2 = np(1 - p).$$

Coin Toss

The binomial distribution with $p = 0.5$ describes the probability of getting k heads in n tosses. Thus the expected value of the number of heads is $\frac{n}{2}$, , and the variance is $\frac{n}{4}$.

Fair Die

A six-sided fair die can be modelled with a discrete random variable with outcomes 1 through 6, each with equal probability $p = 0.5$. The expected value is $\dfrac{1+2+3+4+5+6}{6} = 3.5$. Therefore, the variance can be computed to be

$$\sum_{i=1}^{6} \tfrac{1}{6}(i-3.5)^2 = \tfrac{1}{6}\sum_{i=1}^{6}(i-3.5)^2 = \tfrac{1}{6}\left((-2.5)^2 + (-1.5)^2 + (-0.5)^2 + 0.5^2 + 1.5^2 + 2.5^2\right) = \tfrac{1}{6}\cdot 17.50 = \tfrac{35}{12} \approx 2.92$$

The general formula for the variance of the outcome X of a die of n sides is

$$\sigma^2 = E(X^2) - (E(X))^2 = \frac{1}{n}\sum_{i=1}^{n} i^2 - \left(\frac{1}{n}\sum_{i=1}^{n} i\right)^2$$

$$= \tfrac{1}{6}(n+1)(2n+1) - \tfrac{1}{4}(n+1)^2$$

$$= \frac{n^2-1}{12}.$$

Properties

Basic Properties

Variance is non-negative because the squares are positive or zero.

$$\mathrm{Var}(X) \geq 0.$$

The variance of a constant random variable is zero, and if the variance of a variable in a data set is 0, then all the entries have the same value.

$$P(X = a) = 1 \Leftrightarrow \mathrm{Var}(X) = 0.$$

Variance is invariant with respect to changes in a location parameter. That is, if a constant is added to all values of the variable, the variance is unchanged.

$$\mathrm{Var}(X + a) = \mathrm{Var}(X).$$

If all values are scaled by a constant, the variance is scaled by the square of that constant.

$$\mathrm{Var}(aX) = a^2 \, \mathrm{Var}(X).$$

The variance of a sum of two random variables is given by:

$$\mathrm{Var}(aX+bY)=a^2\,\mathrm{Var}(X)+b^2\,\mathrm{Var}(Y)+2ab\mathrm{Cov}(X,Y),$$

$$\mathrm{Var}(aX-bY)=a^2\,\mathrm{Var}(X)+b^2\,\mathrm{Var}(Y)-2ab\mathrm{Cov}(X,Y),$$

where Cov(., .) is the covariance. In general we have for the sum of N random variables $\{X_1,\ldots,X_N\}$:

$$\mathrm{Var}\left(\sum_{i=1}^{N}X_i\right)=\sum_{i,j=1}^{N}\mathrm{Cov}(X_i,X_j)=\sum_{i=1}^{N}\mathrm{Var}(X_i)+\sum_{i\neq j}\mathrm{Cov}(X_i,X_j).$$

These results lead to the variance of a linear combination as:

$$\mathrm{Var}\left(\sum_{i=1}^{N}a_iX_i\right)=\sum_{i,j=1}^{N}a_ia_j\,\mathrm{Cov}(X_i,X_j)$$

$$=\sum_{i=1}^{N}a_i^2\,\mathrm{Var}(X_i)+\sum_{i\neq j}a_ia_j\,\mathrm{Cov}(X_i,X_j)$$

$$=\sum_{i=1}^{N}a_i^2\,\mathrm{Var}(X_i)+2\sum_{1\leq i<j\leq N}a_ia_j\,\mathrm{Cov}(X_i,X_j)\,.$$

If the random variables $X_1,\ldots,X_N$ are such that

$$\mathrm{Cov}(X_i,X_j)=0\,,\forall\,(i\neq j),$$

they are said to be uncorrelated. It follows immediately from the expression given earlier that if the random variables $X_1,\ldots,X_N$ are uncorrelated, then the variance of their sum is equal to the sum of their variances, or, expressed symbolically:

$$\mathrm{Var}\left(\sum_{i=1}^{N}X_i\right)=\sum_{i=1}^{N}\mathrm{Var}(X_i).$$

Since independent random variables are always uncorrelated, the equation above holds in particular when the random variables $X_1,\ldots,X_n$ are independent. Thus independence is sufficient but not necessary for the variance of the sum to equal the sum of the variances.

Sum of Uncorrelated Variables (Bienaymé formula)

One reason for the use of the variance in preference to other measures of dispersion is that the variance of the sum (or the difference) of uncorrelated random variables is the sum of their variances:

$$\mathrm{Var}\left(\sum_{i=1}^{n} X_i\right) = \sum_{i=1}^{n} \mathrm{Var}(X_i).$$

This statement is called the Bienaymé formula and was discovered in 1853. It is often made with the stronger condition that the variables are independent, but being uncorrelated suffices. So if all the variables have the same variance σ^2, then, since division by n is a linear transformation, this formula immediately implies that the variance of their mean is

$$\mathrm{Var}\left(\overline{X}\right) = \mathrm{Var}\left(\frac{1}{n}\sum_{i=1}^{n} X_i\right) = \frac{1}{n^2}\sum_{i=1}^{n} \mathrm{Var}\left(X_i\right) = \frac{\sigma^2}{n}.$$

That is, the variance of the mean decreases when n increases. This formula for the variance of the mean is used in the definition of the standard error of the sample mean, which is used in the central limit theorem.

Sum of Correlated Variables

In general, if the variables are correlated, then the variance of their sum is the sum of their covariances:

$$\mathrm{Var}\left(\sum_{i=1}^{n} X_i\right) = \sum_{i=1}^{n}\sum_{j=1}^{n} \mathrm{Cov}(X_i, X_j) = \sum_{i=1}^{n} \mathrm{Var}(X_i) + 2\sum_{1 \le i < j \le n} \mathrm{Cov}(X_i, X_j).$$

(Note: The second equality comes from the fact that $\mathrm{Cov}(X_i, X_i) = \mathrm{Var}(X_i)$.)

Here Cov(., .) is the covariance, which is zero for independent random variables (if it exists). The formula states that the variance of a sum is equal to the sum of all elements in the covariance matrix of the components. This formula is used in the theory of Cronbach's alpha in classical test theory.

So if the variables have equal variance σ^2 and the average correlation of distinct variables is ρ, then the variance of their mean is

$$\mathrm{Var}(\overline{X}) = \frac{\sigma^2}{n} + \frac{n-1}{n}\rho\sigma^2.$$

This implies that the variance of the mean increases with the average of the correlations. In other words, additional correlated observations are not as effective as additional independent observations at reducing the uncertainty of the mean. Moreover, if the variables have unit variance, for example if they are standardized, then this simplifies to

$$\mathrm{Var}(\bar{X}) = \frac{1}{n} + \frac{n-1}{n}\rho.$$

This formula is used in the Spearman–Brown prediction formula of classical test theory. This converges to ρ if n goes to infinity, provided that the average correlation remains constant or converges too. So for the variance of the mean of standardized variables with equal correlations or converging average correlation we have

$$\lim_{n\to\infty} \mathrm{Var}(\bar{X}) = \rho.$$

Therefore, the variance of the mean of a large number of standardized variables is approximately equal to their average correlation. This makes clear that the sample mean of correlated variables does not generally converge to the population mean, even though the Law of large numbers states that the sample mean will converge for independent variables.

Matrix Notation for the Variance of a Linear Combination

Define X as a column vector of n random variables $X_1,\ldots,X_n$, and c as a column vector of n scalars $c_1,\ldots,c_n$. Therefore, $c^T X$ is a linear combination of these random variables, where c^T denotes the transpose of c. Also let Σ be the covariance matrix of X. The variance of $c^T X$ is then given by:

$$\mathrm{Var}(c^T X) = c^T \Sigma c.$$

Weighted Sum of Variables

The scaling property and the Bienaymé formula, along with the property of the covariance $\mathrm{Cov}(aX, bY) = ab\,\mathrm{Cov}(X, Y)$ jointly imply that

$$\mathrm{Var}(aX \pm bY) = a^2\,\mathrm{Var}(X) + b^2\,\mathrm{Var}(Y) \pm 2ab\,\mathrm{Cov}(X, Y).$$

This implies that in a weighted sum of variables, the variable with the largest weight will have a disproportionally large weight in the variance of the total. For example, if X and Y are uncorrelated and the weight of X is two times the weight of Y, then the weight of the variance of X will be four times the weight of the variance of Y.

The expression above can be extended to a weighted sum of multiple variables:

$$\mathrm{Var}\left(\sum_i^n a_i X_i\right) = \sum_{i=1}^n a_i^2\,\mathrm{Var}(X_i) + 2\sum\sum_{1\leq i < j\leq n} a_i a_j\,\mathrm{Cov}(X_i, X_j)$$

Product of Independent variables

If two variables X and Y are independent, the variance of their product is given by

$$\mathrm{Var}(XY) = [E(X)]^2 \, \mathrm{Var}(Y) + [E(Y)]^2 \, \mathrm{Var}(X) + \mathrm{Var}(X) \, \mathrm{Var}(Y).$$

Equivalently, using the basic properties of expectation, it is given by

$$\mathrm{Var}(XY) = E(X^2)E(Y^2) - [E(X)]^2[E(Y)]^2.$$

Product of Correlated Variables

In general, if two variables are correlated, the variance of their product is given by

$$\mathrm{Var}(XY) = E[X^2Y^2] - [E(XY)]^2$$

$$= \mathrm{Cov}(X^2, Y^2) + E(X^2)E(Y^2) - [E(XY)]^2$$

$$= \mathrm{Cov}(X^2, Y^2) + (\mathrm{Var}(X) + [E(X)]^2)(\mathrm{Var}(Y) + [E(Y)]^2) - [\mathrm{Cov}(X,Y) + E(X)E(Y)]^2$$

Decomposition

The general formula for variance decomposition or the law of total variance is: If X and Y are two random variables, and the variance of X exists, then

$$\mathrm{Var}[X] = E_Y(\mathrm{Var}[X\,|\,Y]) + \mathrm{Var}_Y(E[X\,|\,Y]).$$

where $E(X\,|\,Y)$ is the conditional expectation of X given Y, and $\mathrm{Var}(X\,|\,Y)$ is the conditional variance of X given Y. (A more intuitive explanation is that given a particular value of Y, then X follows a distribution with mean $E(X\,|\,Y)$ and variance $\mathrm{Var}(X\,|\,Y)$). As $E(X\,|\,Y)$ is a function of the variable Y, the outer expectation or variance is taken with respect to Y. The above formula tells how to find $\mathrm{Var}(X)$ based on the distributions of these two quantities when Y is allowed to vary.

In particular, if Y is a discrete random variable assuming $y_1, y_2, \ldots, y_n$ with corresponding probability masses $p_1, p_2, \ldots, p_n$, then in the formula for total variance, the first term on the right-hand side becomes

$$E_Y(\mathrm{Var}[X\,|\,Y]) = \sum_{i=1}^{n} p_i \sigma_i^2,$$

where $\sigma_i^2 = \mathrm{Var}[X\,|\,y_i]$. Similarly, the second term on the right-hand side becomes

$$\mathrm{Var}_Y(E[X\,|\,Y]) = \sum_{i=1}^{n} p_i \mu_i^2 - \left(\sum_{i=1}^{n} p_i \mu_i\right)^2 = \sum_{i=1}^{n} p_i \mu_i^2 - \mu^2,$$

where $\mu_i = E[X\,|\,y_i]$ and $\mu = \sum_{i=1}^{n} p_i \mu_i$. Thus the total variance is given by

$$\mu = \sum_{i=1}^{n} p_i \mu_i.$$

A similar formula is applied in analysis of variance, where the corresponding formula is

$$MS_{\text{total}} = MS_{\text{between}} + MS_{\text{within}};$$

here MS refers to the Mean of the Squares. In linear regression analysis the corresponding formula is

$$MS_{\text{total}} = MS_{\text{regression}} + MS_{\text{residual}}.$$

This can also be derived from the additivity of variances, since the total (observed) score is the sum of the predicted score and the error score, where the latter two are uncorrelated.

Similar decompositions are possible for the sum of squared deviations (sum of squares, SS):

$$SS_{\text{total}} = SS_{\text{between}} + SS_{\text{within}},$$
$$SS_{\text{total}} = SS_{\text{regression}} + SS_{\text{residual}}.$$

Formulae for the Variance

A formula often used for deriving the variance of a theoretical distribution is as follows:

$$\text{Var}(X) = \text{E}(X^2) - (\text{E}(X))^2.$$

This will be useful when it is possible to derive formulae for the expected value and for the expected value of the square.

This formula is also sometimes used in connection with the sample variance. While useful for hand calculations, it is not advised for computer calculations as it suffers from catastrophic cancellation if the two components of the equation are similar in magnitude and floating point arithmetic is used.

Calculation from the CDF

The population variance for a non-negative random variable can be expressed in terms of the cumulative distribution function F using

$$2\int_0^{\infty} u(1 - F(u))du - \left(\int_0^{\infty} 1 - F(u)du\right)^2.$$

This expression can be used to calculate the variance in situations where the CDF, but not the density, can be conveniently expressed.

Characteristic Property

The second moment of a random variable attains the minimum value when taken around the first moment (i.e., mean) of the random variable, i.e. $\operatorname{argmin}_m E((X - m)^2) = E(X)$. Conversely, if a continuous function φ satisfies $\operatorname{argmin}_m E(\varphi(X - m)) = E(X)$ for all random variables X, then it is necessarily of the form $\varphi(x) = ax^2 + b$, where $a > 0$. This also holds in the multidimensional case.

Units of Measurement

Unlike expected absolute deviation, the variance of a variable has units that are the square of the units of the variable itself. For example, a variable measured in meters will have a variance measured in meters squared. For this reason, describing data sets via their standard deviation or root mean square deviation is often preferred over using the variance. In the dice example the standard deviation is $\sqrt{2.9} \approx 1.7$, slightly larger than the expected absolute deviation of 1.5.

The standard deviation and the expected absolute deviation can both be used as an indicator of the "spread" of a distribution. The standard deviation is more amenable to algebraic manipulation than the expected absolute deviation, and, together with variance and its generalization covariance, is used frequently in theoretical statistics; however the expected absolute deviation tends to be more robust as it is less sensitive to outliers arising from measurement anomalies or an unduly heavy-tailed distribution.

Approximating the Variance of a Function

The delta method uses second-order Taylor expansions to approximate the variance of a function of one or more random variables: Taylor expansions for the moments of functions of random variables. For example, the approximate variance of a function of one variable is given by

$$\operatorname{Var}\left[f(X)\right] \approx \left(f'(E\left[X\right])\right)^2 \operatorname{Var}\left[X\right]$$

provided that f is twice differentiable and that the mean and variance of X are finite.

Population Variance and Sample Variance

Real-world observations such as the measurements of yesterday's rain throughout the day typically cannot be complete sets of all possible observations that could be made. As such, the variance calculated from the finite set will in general not match the variance that would have been calculated from the full population of possible observations. This means that one estimates the mean and variance that would have been calculated from an omniscient set of observations by using an estimator equation. The estimator is a function of the sample of n observations drawn without observational bias from the whole population of potential observations. In this example that sample would be the set of actual measurements of yesterday's rainfall from available rain gauges within the geography of interest.

The simplest estimators for population mean and population variance are simply the mean and variance of the sample, the sample mean and (uncorrected) sample variance – these are consistent estimators (they converge to the correct value as the number of samples increases), but can be improved. Estimating the population variance by taking the sample's variance is close to optimal in general, but can be improved in two ways. Most simply, the sample variance is computed as an average of squared deviations about the (sample) mean, by dividing by n. However, using values other than n improves the estimator in various ways. Four common values for the denominator are n, $n - 1$, $n + 1$, and $n - 1.5$: n is the simplest (population variance of the sample), $n - 1$ eliminates bias, $n + 1$ minimizes mean squared error for the normal distribution, and $n - 1.5$ mostly eliminates bias in unbiased estimation of standard deviation for the normal distribution.

Firstly, if the omniscient mean is unknown (and is computed as the sample mean), then the sample variance is a biased estimator: it underestimates the variance by a factor of $(n - 1) / n$; correcting by this factor (dividing by $n - 1$ instead of n) is called Bessel's correction. The resulting estimator is unbiased, and is called the (corrected) sample variance or unbiased sample variance. For example, when $n = 1$ the variance of a single observation about the sample mean (itself) is obviously zero regardless of the population variance. If the mean is determined in some other way than from the same samples used to estimate the variance then this bias does not arise and the variance can safely be estimated as that of the samples about the (independently known) mean.

Secondly, the sample variance does not generally minimize mean squared error between sample variance and population variance. Correcting for bias often makes this worse: one can always choose a scale factor that performs better than the corrected sample variance, though the optimal scale factor depends on the excess kurtosis of the population, and introduces bias. This always consists of scaling down the unbiased estimator (dividing by a number larger than $n - 1$), and is a simple example of a shrinkage estimator: one "shrinks" the unbiased estimator towards zero. For the normal distribution, dividing by $n + 1$ (instead of $n - 1$ or n) minimizes mean squared error. The resulting estimator is biased, however, and is known as the biased sample variation.

Population Variance

In general, the *population variance* of a *finite* population of size N with values x_i is given by

$$\sigma^2 = \frac{1}{N}\sum_{i=1}^{N}(x_i-\mu)^2 = \frac{1}{N}\sum_{i=1}^{N}\left(x_i^2-2\mu x_i+\mu^2\right) = \left(\frac{1}{N}\sum_{i=1}^{N}x_i^2\right)-2\mu\left(\frac{1}{N}\sum_{i=1}^{N}x_i\right)+\mu^2 = \left(\frac{1}{N}\sum_{i=1}^{N}x_i^2\right)-\mu^2$$

where the population mean is

$$\mu = \frac{1}{N}\sum_{i=1}^{N}x_i \cdot$$

The population variance can also be computed using

$$\sigma^2 = \frac{1}{N^2}\sum_{i<j}\left(x_i-x_j\right)^2 = \frac{1}{2N^2}\sum_{i,j=1}^{N}\left(x_i-x_j\right)^2$$

This is true because

$$\frac{1}{2N^2}\sum_{i,j=1}^{N}\left(x_i-x_j\right)^2 = \frac{1}{2N^2}\sum_{i,j=1}^{N}\left(x_i^2-2x_ix_j+x_j^2\right)$$

$$= \frac{1}{2N}\sum_{j=1}^{N}\left(\frac{1}{N}\sum_{i=1}^{N}x_i^2\right)-\left(\frac{1}{N}\sum_{i=1}^{N}x_i^2\right)\left(\frac{1}{N}\sum_{j=1}^{N}x_j^2\right)+\frac{1}{2N}\sum_{i=1}^{N}\left(\frac{1}{N}\sum_{j=1}^{N}x_j^2\right)$$

$$= \frac{1}{2}\left(\sigma^2+\mu^2\right)-\mu^2+\frac{1}{2}\left(\sigma^2+\mu^2\right) = \sigma^2$$

The population variance therefore matches the variance of the generating probability distribution. In this sense, the concept of population can be extended to continuous random variables with infinite populations.

Sample Variance

In many practical situations, the true variance of a population is not known *a priori* and must be computed somehow. When dealing with extremely large populations, it is not possible to count every object in the population, so the computation must be performed on a sample of the population. Sample variance can also be applied to the estimation of the variance of a continuous distribution from a sample of that distribution.

We take a sample with replacement of n values $y_1, ..., y_n$ from the population, where $n < N$, and estimate the variance on the basis of this sample. Directly taking the variance of the sample data gives the average of the squared deviations:

$$\sigma_y^2 = \frac{1}{n}\sum_{i=1}^{n}\left(y_i-\bar{y}\right)^2 = \left(\frac{1}{n}\sum_{i=1}^{n}y_i^2\right)-\bar{y}^2 = \frac{1}{n^2}\sum_{i<j}\left(y_i-y_j\right)^2.$$

Here, $\bar{}$ denotes the sample mean:

$$\bar{y} = \frac{1}{n}\sum_{i=1}^{n} y_i.$$

Since the y_i are selected randomly, both $\bar{y}$ and σ_y^2 are random variables. Their expected values can be evaluated by summing over the ensemble of all possible samples $\{y_i\}$ from the population. For σ_y^2 this gives:

$$E[\sigma_y^2] = E\left[\frac{1}{n}\sum_{i=1}^{n}\left(y_i - \frac{1}{n}\sum_{j=1}^{n} y_j\right)^2\right]$$

$$= \frac{1}{n}\sum_{i=1}^{n} E\left[y_i^2 - \frac{2}{n}y_i\sum_{j=1}^{n} y_j + \frac{1}{n^2}\sum_{j=1}^{n} y_j\sum_{k=1}^{n} y_k\right]$$

$$= \frac{1}{n}\sum_{i=1}^{n}\left[\frac{n-2}{n}E[y_i^2] - \frac{2}{n}\sum_{j\neq i} E[y_i y_j] + \frac{1}{n^2}\sum_{j=1}^{n}\sum_{k\neq j} E[y_j y_k] + \frac{1}{n^2}\sum_{j=1}^{n} E[y_j^2]\right]$$

$$= \frac{1}{n}\sum_{i=1}^{n}\left[\frac{n-2}{n}(\sigma^2 + \mu^2) - \frac{2}{n}(n-1)\mu^2 + \frac{1}{n^2}n(n-1)\mu^2 + \frac{1}{n}(\sigma^2 + \mu^2)\right]$$

$$= \frac{n-1}{n}\sigma^2.$$

Hence σ_y^2 gives an estimate of the population variance that is biased by a factor

of $\dfrac{n-1}{n}$. For this reason, σ_y^2 is referred to as the *biased sample variance*. Correcting for this bias yields the *unbiased sample variance*:

$$s^2 = \frac{n}{n-1}\sigma_y^2 = \frac{n}{n-1}\left(\frac{1}{n}\sum_{i=1}^{n}(y_i - \bar{y})^2\right) = \frac{1}{n-1}\sum_{i=1}^{n}(y_i - \bar{y})^2$$

Either estimator may be simply referred to as the *sample variance* when the version can be determined by context. The same proof is also applicable for samples taken from a continuous probability distribution.

The use of the term $n-1$ is called Bessel's correction, and it is also used in sample covariance and the sample standard deviation (the square root of variance). The square root is a concave function and thus introduces negative bias (by Jensen's inequality), which depends on the distribution, and thus the corrected sample standard deviation (using Bessel's correction) is biased. The unbiased estimation of standard deviation is a technically involved problem, though for the normal distribution using the term $n-1.5$ yields an almost unbiased estimator.

The unbiased sample variance is a U-statistic for the function $f(y_1, y_2) = (y_1 - y_2)^2/2$, meaning that it is obtained by averaging a 2-sample statistic over 2-element subsets of the population.

Distribution of the Sample Variance

Being a function of random variables, the sample variance is itself a random variable, and it is natural to study its distribution. In the case that y_i are independent observations from a normal distribution, Cochran's theorem shows that s^2 follows a scaled chi-squared distribution:

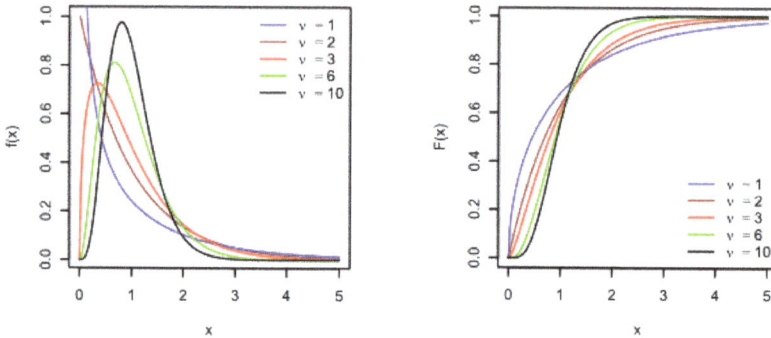

Distribution and cumulative distribution of s^2/σ^2, for various values of $v = n - 1$, when the y_i are independent normally distributed.

$$(n-1)\frac{s^2}{\sigma^2} \sim \chi^2_{n-1}.$$

As a direct consequence, it follows that

$$E(s^2) = E\left(\frac{\sigma^2}{n-1}\chi^2_{n-1}\right) = \sigma^2,$$

and

$$\text{Var}[s^2] = \text{Var}\left(\frac{\sigma^2}{n-1}\chi^2_{n-1}\right) = \frac{\sigma^4}{(n-1)^2}\text{Var}\left(\chi^2_{n-1}\right) = \frac{2\sigma^4}{n-1}.$$

If the y_i are independent and identically distributed, but not necessarily normally distributed, then

$$E[s^2] = \sigma^2, \quad \text{Var}[s^2] = \frac{\sigma^4}{n}\left((\kappa-1)+\frac{2}{n-1}\right) = \frac{1}{n}\left(\mu_4 - \frac{n-3}{n-1}\sigma^4\right),$$

where κ is the kurtosis of the distribution and μ_4 is the fourth central moment.

If the conditions of the law of large numbers hold for the squared observations, s^2 is a consistent estimator of σ^2. One can see indeed that the variance of the estimator tends asymptotically to zero.

Samuelson's Inequality

Samuelson's inequality is a result that states bounds on the values that individual observations in a sample can take, given that the sample mean and (biased) variance have been calculated. Values must lie within the limits $\bar{y} \pm \sigma_y (n-1)^{1/2}$.

Relations with the Harmonic and Arithmetic Means

It has been shown that for a sample $\{y_i\}$ of real numbers,

$$\sigma_y^2 \leq 2 y_{max}(A - H),$$

where y_{max} is the maximum of the sample, A is the arithmetic mean, H is the harmonic mean of the sample and σ_y^2 is the (biased) variance of the sample.

This bound has been improved, and it is known that variance is bounded by

$$\sigma_y^2 \leq \frac{y_{max}(A - H)(y_{max} - A)}{y_{max} - H},$$

$$\sigma_y^2 \geq \frac{y_{min}(A - H)(A - y_{min})}{H - y_{min}},$$

where y_{min} is the minimum of the sample.

Tests of Equality of Variances

Testing for the equality of two or more variances is difficult. The F test and chi square tests are both adversely affected by non-normality and are not recommended for this purpose.

Several non parametric tests have been proposed: these include the Barton–David–Ansari–Freund–Siegel–Tukey test, the Capon test, Mood test, the Klotz test and the Sukhatme test. The Sukhatme test applies to two variances and requires that both medians be known and equal to zero. The Mood, Klotz, Capon and Barton–David–Ansari–Freund–Siegel–Tukey tests also apply to two variances. They allow the median to be unknown but do require that the two medians are equal.

The Lehmann test is a parametric test of two variances. Of this test there are several variants known. Other tests of the equality of variances include the Box test, the Box–Anderson test and the Moses test.

Resampling methods, which include the bootstrap and the jackknife, may be used to test the equality of variances.

History

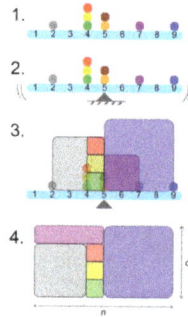

Geometric visualisation of the variance of an arbitrary distribution (2, 4, 4, 4, 5, 5, 7, 9): 1. A frequency distribution is constructed. 2. The centroid of the distribution gives its mean. 3. A square with sides equal to the difference of each value from the mean is formed for each value. 4. Arranging the squares into a rectangle with one side equal to the number of values, n, results in the other side being the distribution's variance, σ^2.

The term *variance* was first introduced by Ronald Fisher in his 1918 paper *The Correlation Between Relatives on the Supposition of Mendelian Inheritance*:

The great body of available statistics show us that the deviations of a human measurement from its mean follow very closely the Normal Law of Errors, and, therefore, that the variability may be uniformly measured by the standard deviation corresponding to the square root of the mean square error. When there are two independent causes of variability capable of producing in an otherwise uniform population distributions with standard deviations σ_1 and σ_2, it is found that the distribution, when both causes act together, has a standard deviation $\sqrt{\sigma_1^2 + \sigma_2^2}$... It is therefore desirable in analysing the causes of variability to deal with the square of the standard deviation as the measure of variability. We shall term this quantity the Variance...

Moment of Inertia

The variance of a probability distribution is analogous to the moment of inertia in classical mechanics of a corresponding mass distribution along a line, with respect to rotation about its center of mass. It is because of this analogy that such things as the variance are called *moments* of probability distributions. The covariance matrix is related to the moment of inertia tensor for multivariate distributions. The moment of inertia of a cloud of n points with a covariance matrix of Σ is given by

$$I = n(1_{3\times3}\,\mathrm{tr}(\Sigma) - \Sigma).$$

This difference between moment of inertia in physics and in statistics is clear for points that are gathered along a line. Suppose many points are close to the x axis and distributed along it. The covariance matrix might look like

$$\Sigma = \begin{bmatrix} 10 & 0 & 0 \\ 0 & 0.1 & 0 \\ 0 & 0 & 0.1 \end{bmatrix}.$$

That is, there is the most variance in the x direction. Physicists would consider this to have a low moment *about* the x axis so the moment-of-inertia tensor is

$$i = n \begin{bmatrix} 0.2 & 0 & 0 \\ 0 & 10.1 & 0 \\ 0 & 0 & 10.1 \end{bmatrix}.$$

Semivariance

The *semivariance* is calculated in the same manner as the variance but only those observations that fall below the mean are included in the calculation. It is sometimes described as a measure of downside risk in an investments context. For skewed distributions, the semivariance can provide additional information that a variance does not.

Generalizations

If x is a scalar complex-valued random variable, with values in $\mathbb{C}$, then its variance is $E((x-\mu)(x-\mu)^*)$, where x^* is the complex conjugate of x. This variance is a real scalar.

If X is a vector-valued random variable, with values in $\mathbb{R}^n$, and thought of as a column vector, then the natural generalization of variance is $E((X-\mu)(X-\mu)^T)$, where $\mu = E(X)$ and X^T is the transpose of X, and so is a row vector. The result is a positive semi-definite square matrix, commonly referred to as the variance-covariance matrix.

If X is a vector- and complex-valued random variable, with values in $\mathbb{C}^n$, then the generalization of its variance is $E((X-\mu)(X-\mu)^\dagger)$, where $X^\dagger$ is the conjugate transpose of X. This matrix is also positive semi-definite and square.

Standard Deviation

A plot of a normal distribution (or bell-shaped curve) where each band has a width of 1 standard deviation –

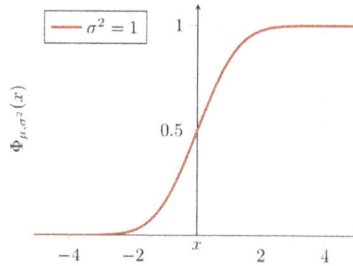

Cumulative probability of a normal distribution with expected value 0 and standard deviation 1.

In statistics, the standard deviation (SD, also represented by the Greek letter sigma σ or the Latin letter s) is a measure that is used to quantify the amount of variation or dispersion of a set of data values. A low standard deviation indicates that the data points tend to be close to the mean (also called the expected value) of the set, while a high standard deviation indicates that the data points are spread out over a wider range of values.

The standard deviation of a random variable, statistical population, data set, or probability distribution is the square root of its variance. It is algebraically simpler, though in practice less robust, than the average absolute deviation. A useful property of the standard deviation is that, unlike the variance, it is expressed in the same units as the data. There are also other measures of deviation from the norm, including mean absolute deviation, which provide different mathematical properties from standard deviation.

In addition to expressing the variability of a population, the standard deviation is commonly used to measure confidence in statistical conclusions. For example, the margin of error in polling data is determined by calculating the expected standard deviation in the results if the same poll were to be conducted multiple times. This derivation of a standard deviation is often called the "standard error" of the estimate or "standard error of the mean" when referring to a mean. It is computed as the standard deviation of all the means that would be computed from that population if an infinite number of samples were drawn and a mean for each sample were computed. It is very important to note that the standard deviation of a population and the standard error of a statistic derived from that population (such as the mean) are quite different but related (related by the inverse of the square root of the number of observations). The reported margin of error of a poll is computed from the standard error of the mean (or alternatively from the product of the standard deviation of the population and the inverse of the square root of the sample size, which is the same thing) and is typically about twice the standard deviation—the half-width of a 95 percent confidence interval. In science, researchers commonly report the standard deviation of experimental data, and only effects that fall much farther than two standard deviations away from what would have been expected are considered statistically significant—normal random error or variation in the measurements is in this way distinguished from likely genuine effects or associations. The standard deviation is also important in finance, where the standard deviation on the rate of return on an investment is a measure of the volatility of the investment.

When only a sample of data from a population is available, the term standard deviation of the sample or sample standard deviation can refer to either the above-mentioned quantity as applied to those data or to a modified quantity that is an unbiased estimate of the population standard deviation (the standard deviation of the entire population).

Basic Examples

For a finite set of numbers, the standard deviation is found by taking the square root of the average of the squared deviations of the values from their average value. For example, the marks of a class of eight students (that is, a population) are the following eight values:

$$2, 4, 4, 4, 5, 5, 7, 9.$$

These eight data points have the mean (average) of 5:

$$\frac{2+4+4+4+5+5+7+9}{8} = 5.$$

First, calculate the deviations of each data point from the mean, and square the result of each:

$$(2-5)^2 = (-3)^2 = 9 \qquad (5-5)^2 = 0^2 = 0$$
$$(4-5)^2 = (-1)^2 = 1 \qquad (5-5)^2 = 0^2 = 0$$
$$(4-5)^2 = (-1)^2 = 1 \qquad (7-5)^2 = 2^2 = 4$$
$$(4-5)^2 = (-1)^2 = 1 \qquad (9-5)^2 = 4^2 = 16.$$

The variance is the mean of these values:

$$\frac{9+1+1+1+0+0+4+16}{8} = 4.$$

and the *population* standard deviation is equal to the square root of the variance:

$$\sqrt{4} = 2.$$

This formula is valid only if the eight values with which we began form the complete population. If the values instead were a random sample drawn from some large parent population (for example, they were 8 marks randomly and independently chosen from a class of 2 million), then one often divides by 7 (which is $n-1$) instead of 8 (which is n) in the denominator of the last formula. In that the result would be called the *sample* standard deviation. Dividing by $n-1$ rather than by n gives an unbiased estimate of the standard deviation of the larger parent population. This is known as *Bessel's correction*.

As a slightly more complicated real-life example, the average height for adult men in the United States is about 70 inches, with a standard deviation of around 3 inches.

This means that most men (about 68%, assuming a normal distribution) have a height within 3 inches of the mean (67–73 inches) – one standard deviation – and almost all men (about 95%) have a height within 6 inches of the mean (64–76 inches) – two standard deviations. If the standard deviation were zero, then all men would be exactly 70 inches tall. If the standard deviation were 20 inches, then men would have much more variable heights, with a typical range of about 50–90 inches. Three standard deviations account for 99.7% of the sample population being studied, assuming the distribution is normal (bell-shaped).

Definition of Population Values

Let X be a random variable with mean value μ:

$$\mathrm{E}[X] = \mu.$$

Here the operator E denotes the average or expected value of X. Then the standard deviation of X is the quantity

$$\sigma = \sqrt{\mathrm{E}[(X - \mu)^2]}$$

$$= \sqrt{\mathrm{E}[X^2] + \mathrm{E}[-2\mu X] + \mathrm{E}[\mu^2]} = \sqrt{\mathrm{E}[X^2] - 2\mu \mathrm{E}[X] + \mu^2}$$

$$= \sqrt{\mathrm{E}[X^2] - 2\mu^2 + \mu^2} = \sqrt{\mathrm{E}[X^2] - \mu^2}$$

$$= \sqrt{\mathrm{E}[X^2] - (\mathrm{E}[X])^2}$$

(derived using the properties of expected value).

In other words, the standard deviation σ (sigma) is the square root of the variance of X; i.e., it is the square root of the average value of $(X - \mu)^2$.

The standard deviation of a (univariate) probability distribution is the same as that of a random variable having that distribution. Not all random variables have a standard deviation, since these expected values need not exist. For example, the standard deviation of a random variable that follows a Cauchy distribution is undefined because its expected value μ is undefined.

Discrete Random Variable

In the case where X takes random values from a finite data set $x_1, x_2, ..., x_N$, with each value having the same probability, the standard deviation is

$$\sigma = \sqrt{\frac{1}{N}\left[(x_1 - \mu)^2 + (x_2 - \mu)^2 + \cdots + (x_N - \mu)^2\right]}, \text{ where } \mu = \frac{1}{N}(x_1 + \cdots + x_N),$$

or, using summation notation,

$$\sigma = \sqrt{\frac{1}{N}\sum_{i=1}^{N}(x_i - \mu)^2}, \text{ where } \mu = \frac{1}{N}\sum_{i=1}^{N}x_i.$$

If, instead of having equal probabilities, the values have different probabilities, let x_1 have probability p_1, x_2 have probability p_2, ..., x_N have probability p_N. In this case, the standard deviation will be

$$\sigma = \sqrt{\sum_{i=1}^{N}p_i(x_i - \mu)^2}, \text{ where } \mu = \sum_{i=1}^{N}p_i x_i.$$

Continuous Random Variable

The standard deviation of a continuous real-valued random variable X with probability density function $p(x)$ is

$$\sigma = \sqrt{\int_{X}(x - \mu)^2 p(x)\mathrm{d}x}, \text{ where } \mu = \int_{X}x p(x)\mathrm{d}x,$$

and where the integrals are definite integrals taken for x ranging over the set of possible values of the random variable X.

In the case of a parametric family of distributions, the standard deviation can be expressed in terms of the parameters. For example, in the case of the log-normal distribution with parameters μ and σ^2, the standard deviation is $[(\exp(\sigma^2) - 1)\exp(2\mu + \sigma^2)]^{1/2}$.

Estimation

One can find the standard deviation of an entire population in cases (such as standardized testing) where every member of a population is sampled. In cases where that cannot be done, the standard deviation σ is estimated by examining a random sample taken from the population and computing a statistic of the sample, which is used as an estimate of the population standard deviation. Such a statistic is called an estimator, and the estimator (or the value of the estimator, namely the estimate) is called a sample standard deviation, and is denoted by s (possibly with modifiers). However, unlike in the case of estimating the population mean, for which the sample mean is a simple estimator with many desirable properties (unbiased, efficient, maximum likelihood), there is no single estimator for the standard deviation with all these properties, and unbiased estimation of standard deviation is a very technically involved problem. Most often, the standard deviation is estimated using the *correct-*

ed *sample standard deviation* (using $N - 1$), defined below, and this is often referred to as the "sample standard deviation", without qualifiers. However, other estimators are better in other respects: the uncorrected estimator (using N) yields lower mean squared error, while using $N - 1.5$ (for the normal distribution) almost completely eliminates bias.

Uncorrected Sample Standard Deviation

Firstly, the formula for the *population* standard deviation (of a finite population) can be applied to the sample, using the size of the sample as the size of the population (though the actual population size from which the sample is drawn may be much larger). This estimator, denoted by s_N, is known as the uncorrected sample standard deviation, or sometimes the standard deviation of the sample (considered as the entire population), and is defined as follows:

$$s_N = \sqrt{\frac{1}{N}\sum_{i=1}^{N}(x_i - \bar{x})^2},$$

where $\{x_1, x_2, \ldots, x_N\}$ are the observed values of the sample items and $\bar{x}$ is the mean value of these observations, while the denominator N stands for the size of the sample: this is the square root of the sample variance, which is the average of the squared deviations about the sample mean.

This is a consistent estimator (it converges in probability to the population value as the number of samples goes to infinity), and is the maximum-likelihood estimate when the population is normally distributed. However, this is a biased estimator, as the estimates are generally too low. The bias decreases as sample size grows, dropping off as $1/n$, and thus is most significant for small or moderate sample sizes; for $n > 75$ the bias is below 1%. Thus for very large sample sizes, the uncorrected sample standard deviation is generally acceptable. This estimator also has a uniformly smaller mean squared error than the corrected sample standard deviation.

Corrected sample standard deviation

If the *biased sample variance* (the second central moment of the sample, which is a downward-biased estimate of the population variance) is used to compute an estimate of the population's standard deviation, the result is

$$s_N = \sqrt{\frac{1}{N}\sum_{i=1}^{N}(x_i - \bar{x})^2}.$$

Here taking the square root introduces further downward bias, by Jensen's inequality, due to the square root being a concave function. The bias in the variance is easily corrected, but the bias from the square root is more difficult to correct, and depends on the distribution in question.

An unbiased estimator for the *variance* is given by applying Bessel's correction, using $N-1$ instead of N to yield the *unbiased sample variance,* denoted s^2:

$$s^2 = \frac{1}{N-1} \sum_{i=1}^{N} (x_i - \overline{x})^2.$$

This estimator is unbiased if the variance exists and the sample values are drawn independently with replacement. $N-1$ corresponds to the number of degrees of freedom in the vector of deviations from the mean, $(x_1 - \overline{x}, \ldots, x_n - \overline{x})$.

Taking square roots reintroduces bias (because the square root is a nonlinear function, which does not commute with the expectation), yielding the corrected sample standard deviation, denoted by s:

$$s = \sqrt{\frac{1}{N-1} \sum_{i=1}^{N} (x_i - \overline{x})^2}.$$

As explained above, while s^2 is an unbiased estimator for the population variance, s is still a biased estimator for the population standard deviation, though markedly less biased than the uncorrected sample standard deviation. The bias is still significant for small samples (N less than 10), and also drops off as $1/N$ as sample size increases. This estimator is commonly used and generally known simply as the "sample standard deviation".

Unbiased Sample Standard Deviation

For unbiased estimation of standard deviation, there is no formula that works across all distributions, unlike for mean and variance. Instead, s is used as a basis, and is scaled by a correction factor to produce an unbiased estimate. For the normal distribution, an unbiased estimator is given by s/c_4, where the correction factor (which depends on N) is given in terms of the Gamma function, and equals:

$$c_4(N) = \sqrt{\frac{2}{N-1}} \, \frac{\Gamma\left(\dfrac{N}{2}\right)}{\Gamma\left(\dfrac{N-1}{2}\right)}.$$

This arises because the sampling distribution of the sample standard deviation follows a (scaled) chi distribution, and the correction factor is the mean of the chi distribution.

An approximation can be given by replacing $N-1$ with $N-1.5$, yielding:

$$\hat{\sigma} = \sqrt{\frac{1}{N-1.5} \sum_{i=1}^{n} (x_i - \bar{x})^2},$$

The error in this approximation decays quadratically (as $1/N^2$), and it is suited for all but the smallest samples or highest precision: for n = 3 the bias is equal to 1.3%, and for n = 9 the bias is already less than 0.1%.

For other distributions, the correct formula depends on the distribution, but a rule of thumb is to use the further refinement of the approximation:

$$\hat{\sigma} = \sqrt{\frac{1}{n-1.5-\frac{1}{4}\gamma_2} \sum_{i=1}^{n} (x_i - \bar{x})^2},$$

where γ_2 denotes the population excess kurtosis. The excess kurtosis may be either known beforehand for certain distributions, or estimated from the data.

Confidence Interval of a Sampled Standard Deviation

The standard deviation we obtain by sampling a distribution is itself not absolutely accurate, both for mathematical reasons (explained here by the confidence interval) and for practical reasons of measurement (measurement error). The mathematical effect can be described by the confidence interval or CI. To show how a larger sample will make the confidence interval narrower, consider the following examples: A small population of $N = 2$ has only 1 degree of freedom for estimating the standard deviation. The result is that a 95% CI of the SD runs from 0.45*SD to 31.9*SD; the factors here are as follows:

$$\Pr\{q_{\alpha/2} < ks^2 / \sigma^2 < q_{1-\alpha/2}\} = 1-\alpha,$$

where q_p is the p-th quantile of the chi-square distribution with k degrees of freedom, and $1-\alpha$ is the confidence level. This is equivalent to the following:

$$\Pr\{(ks^2)/q_{1-\alpha/2} < \sigma^2 < (ks^2)/q_{\alpha/2}\} = 1-\alpha.$$

With $k = 1$, $q_{0.025} = 0.000982$ and $q_{0.975} = 5.024$. The reciprocals of the square roots of these two numbers give us the factors 0.45 and 31.9 given above.

A larger population of $N = 10$ has 9 degrees of freedom for estimating the standard deviation. The same computations as above give us in this case a 95% CI running from 0.69*SD to 1.83*SD. So even with a sample population of 10, the actual SD can still be almost a factor 2 higher than the sampled SD. For a sample population N=100, this is down to 0.88*SD to 1.16*SD. To be more certain that the sampled SD is close to the actual SD we need to sample a large number of points.

These same formulae can be used to obtain confidence intervals on the variance of residuals from a least squares fit under standard normal theory, where k is now the number of degrees of freedom for error.

Identities and Mathematical Properties

The standard deviation is invariant under changes in location, and scales directly with the scale of the random variable. Thus, for a constant c and random variables X and Y:

$$\sigma(c) = 0$$

$$\sigma(X + c) = \sigma(X),$$

$$\sigma(cX) = |c| \, \sigma(X).$$

The standard deviation of the sum of two random variables can be related to their individual standard deviations and the covariance between them:

$$\sigma(X + Y) = \sqrt{\mathrm{var}(X) + \mathrm{var}(Y) + 2\mathrm{cov}(X, Y)}.$$

where $\mathrm{var} = \sigma^2$ and cov stand for variance and covariance, respectively.

The calculation of the sum of squared deviations can be related to moments calculated directly from the data. In the following formula, the letter E is interpreted to mean expected value, i.e., mean.

$$\sigma(X) = \sqrt{E[(X - E(X))^2]} = \sqrt{E[X^2] - (E[X])^2}.$$

The sample standard deviation can be computed as:

$$\sigma(X) = \sqrt{\frac{N}{N-1}} \sqrt{E[(X - E(X))^2]}.$$

For a finite population with equal probabilities at all points, we have

$$\sqrt{\frac{1}{N} \sum_{i=1}^{N} (x_i - \overline{x})^2} = \sqrt{\frac{1}{N} \left(\sum_{i=1}^{N} x_i^2 \right) - \overline{x}^2} = \sqrt{\left(\frac{1}{N} \sum_{i=1}^{N} x_i^2 \right) - \left(\frac{1}{N} \sum_{i=1}^{N} x_i \right)^2}.$$

This means that the standard deviation is equal to the square root of the difference between the average of the squares of the values and the square of the average value.

Interpretation and Application

A large standard deviation indicates that the data points can spread far from the mean and a small standard deviation indicates that they are clustered closely around the mean.

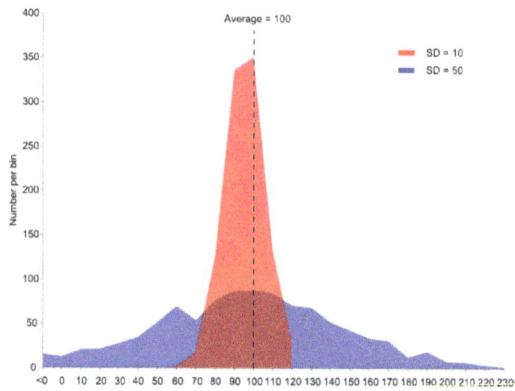

Example of samples from two populations with the same mean but different standard deviations. Red population has mean 100 and SD 10; blue population has mean 100 and SD 50.

For example, each of the three populations $\{0, 0, 14, 14\}$, $\{0, 6, 8, 14\}$ and $\{6, 6, 8, 8\}$ has a mean of 7. Their standard deviations are 7, 5, and 1, respectively. The third population has a much smaller standard deviation than the other two because its values are all close to 7. It will have the same units as the data points themselves. If, for instance, the data set $\{0, 6, 8, 14\}$ represents the ages of a population of four siblings in years, the standard deviation is 5 years. As another example, the population $\{1000, 1006, 1008, 1014\}$ may represent the distances traveled by four athletes, measured in meters. It has a mean of 1007 meters, and a standard deviation of 5 meters.

Standard deviation may serve as a measure of uncertainty. In physical science, for example, the reported standard deviation of a group of repeated measurements gives the precision of those measurements. When deciding whether measurements agree with a theoretical prediction, the standard deviation of those measurements is of crucial importance: if the mean of the measurements is too far away from the prediction (with the distance measured in standard deviations), then the theory being tested probably needs to be revised. This makes sense since they fall outside the range of values that could reasonably be expected to occur, if the prediction were correct and the standard deviation appropriately quantified.

While the standard deviation does measure how far typical values tend to be from the mean, other measures are available. An example is the mean absolute deviation, which might be considered a more direct measure of average distance, compared to the root mean square distance inherent in the standard deviation.

Application Examples

The practical value of understanding the standard deviation of a set of values is in appreciating how much variation there is from the average (mean).

Experiment, Industrial and Hypothesis Testing

Standard deviation is often used to compare real-world data against a model to test the model. For example, in industrial applications the weight of products coming off a production line may need to comply with a legally required value. By weighing some fraction of the products an average weight can be found, which will always be slightly different to the long-term average. By using standard deviations, a minimum and maximum value can be calculated that the averaged weight will be within some very high percentage of the time (99.9% or more). If it falls outside the range then the production process may need to be corrected. Statistical tests such as these are particularly important when the testing is relatively expensive. For example, if the product needs to be opened and drained and weighed, or if the product was otherwise used up by the test.

In experimental science, a theoretical model of reality is used. Particle physics conventionally uses a standard of "5 sigma" for the declaration of a discovery. A five-sigma level translates to one chance in 3.5 million that a random fluctuation would yield the result. This level of certainty was required in order to assert that a particle consistent with the Higgs boson had been discovered in two independent experiments at CERN, and this was also the significance level leading to the declaration of the first detection of gravitational waves.

Weather

As a simple example, consider the average daily maximum temperatures for two cities, one inland and one on the coast. It is helpful to understand that the range of daily maximum temperatures for cities near the coast is smaller than for cities inland. Thus, while these two cities may each have the same average maximum temperature, the standard deviation of the daily maximum temperature for the coastal city will be less than that of the inland city as, on any particular day, the actual maximum temperature is more likely to be farther from the average maximum temperature for the inland city than for the coastal one.

Finance

In finance, standard deviation is often used as a measure of the risk associated with price-fluctuations of a given asset (stocks, bonds, property, etc.), or the risk of a portfolio of assets (actively managed mutual funds, index mutual funds, or ETFs). Risk is an important factor in determining how to efficiently manage a portfolio of investments because it determines the variation in returns on the asset and/or portfolio and gives investors a mathematical basis for investment decisions (known as mean-variance optimization). The fundamental concept of risk is that as it increases, the expected return on an investment should increase as well, an increase known as the risk premium. In other words, investors should expect a higher return on an investment when that investment carries a higher level of risk or uncertainty. When evaluating investments, investors should estimate both the expected return and the uncertainty of future returns. Standard deviation provides a quantified estimate of the uncertainty of future returns.

For example, assume an investor had to choose between two stocks. Stock A over the past 20 years had an average return of 10 percent, with a standard deviation of 20 percentage points (pp) and Stock B, over the same period, had average returns of 12 percent but a higher standard deviation of 30 pp. On the basis of risk and return, an investor may decide that Stock A is the safer choice, because Stock B's additional two percentage points of return is not worth the additional 10 pp standard deviation (greater risk or uncertainty of the expected return). Stock B is likely to fall short of the initial investment (but also to exceed the initial investment) more often than Stock A under the same circumstances, and is estimated to return only two percent more on average. In this example, Stock A is expected to earn about 10 percent, plus or minus 20 pp (a range of 30 percent to −10 percent), about two-thirds of the future year returns. When considering more extreme possible returns or outcomes in future, an investor should expect results of as much as 10 percent plus or minus 60 pp, or a range from 70 percent to −50 percent, which includes outcomes for three standard deviations from the average return (about 99.7 percent of probable returns).

Calculating the average (or arithmetic mean) of the return of a security over a given period will generate the expected return of the asset. For each period, subtracting the expected return from the actual return results in the difference from the mean. Squaring the difference in each period and taking the average gives the overall variance of the return of the asset. The larger the variance, the greater risk the security carries. Finding the square root of this variance will give the standard deviation of the investment tool in question.

Population standard deviation is used to set the width of Bollinger Bands, a widely adopted technical analysis tool. For example, the upper Bollinger Band is given as $x + n\sigma_x$. The most commonly used value for n is 2; there is about a five percent chance of going outside, assuming a normal distribution of returns.

Financial time series are known to be non-stationary series, whereas the statistical calculations above, such as standard deviation, apply only to stationary series. To apply the above statistical tools to non-stationary series, the series first must be transformed to a stationary series, enabling use of statistical tools that now have a valid basis from which to work.

Geometric Interpretation

To gain some geometric insights and clarification, we will start with a population of three values, x_1, x_2, x_3. This defines a point $P = (x_1, x_2, x_3)$ in R^3. Consider the line $L = \{(r, r, r) : r \in R\}$. This is the "main diagonal" going through the origin. If our three given values were all equal, then the standard deviation would be zero and P would lie on L. So it is not unreasonable to assume that the standard deviation is related to the *distance* of P to L. That is indeed the case. To move orthogonally from L to the point P, one begins at the point:

$$M = (\overline{x}, \overline{x}, \overline{x})$$

whose coordinates are the mean of the values we started out with.

A little algebra shows that the distance between P and M (which is the same as the orthogonal distance between P and the line L) $\sqrt{\sum_i (x_i - \overline{x})^2}$ is equal to the standard deviation of the vector x_1, x_2, x_3, multiplied by the square root of the number of dimensions of the vector (3 in this case.)

Chebyshev's Inequality

An observation is rarely more than a few standard deviations away from the mean. Chebyshev's inequality ensures that, for all distributions for which the standard deviation is defined, the amount of data within a number of standard deviations of the mean is at least as much as given in the following table.

Distance from mean	Minimum population
$\sqrt{2}\sigma$	50%
2σ	75%
3σ	89%
4σ	94%
5σ	96%
6σ	97%
$k\sigma$	$1 - \dfrac{1}{k^2}$
$\dfrac{1}{\sqrt{1-l}}\sigma$	l

Rules for Normally Distributed data

Dark blue is one standard deviation on either side of the mean. For the normal distribution, this accounts for 68.27 percent of the set; while two standard deviations from the mean (medium and dark blue) account for 95.45 percent; three standard deviations (light, medium, and dark blue) account for 99.73 percent; and four standard deviations account for 99.994 percent. The two points of the curve that are one standard deviation from the mean are also the inflection points.

The central limit theorem says that the distribution of an average of many independent, identically distributed random variables tends toward the famous bell-shaped normal distribution with a probability density function of:

$$f(x; \mu, \sigma^2) = \frac{1}{\sigma\sqrt{2\pi}} e^{-\frac{1}{2}\left(\frac{x-\mu}{\sigma}\right)^2}$$

where μ is the expected value of the random variables, σ equals their distribution's standard deviation divided by $n^{1/2}$, and n is the number of random variables. The standard deviation therefore is simply a scaling variable that adjusts how broad the curve will be, though it also appears in the normalizing constant.

If a data distribution is approximately normal, then the proportion of data values within z standard deviations of the mean is defined by:

$$\text{Proportion} = \text{erf}\left(\frac{z}{\sqrt{2}}\right)$$

where erf is the error function. The proportion that is less than or equal to a number, x, is given by the cumulative distribution function:

$$\text{Proportion} \le x = \frac{1}{2}\left[1 + \text{erf}\left(\frac{x-\mu}{\sigma\sqrt{2}}\right)\right] = \frac{1}{2}\left[1 + \text{erf}\left(\frac{z}{\sqrt{2}}\right)\right].$$

If a data distribution is approximately normal then about 68 percent of the data values are within one standard deviation of the mean (mathematically, $\mu \pm \sigma$, where μ is the arithmetic mean), about 95 percent are within two standard deviations ($\mu \pm 2\sigma$), and about 99.7 percent lie within three standard deviations ($\mu \pm 3\sigma$). This is known as the 68-95-99.7 rule, or the empirical rule.

For various values of z, the percentage of values expected to lie in and outside the symmetric interval, CI = $(-z\sigma, z\sigma)$, are as follows:

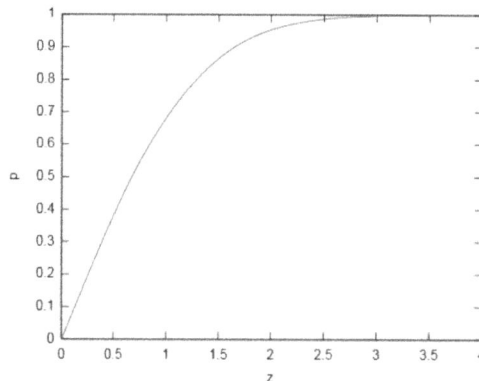

Percentage within(z)

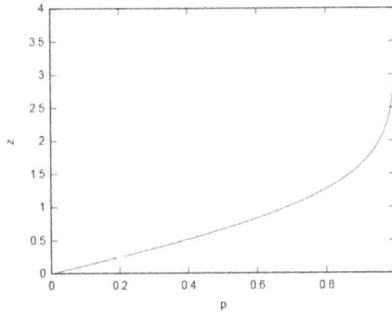

z(Percentage within)

Confidence interval	Proportion within	Proportion without	
	Percentage	Percentage	Fraction
0.674490σ	50%	50%	1 / 2
0.994458σ	68%	32%	1 / 3.125
1σ	68.2689492%	31.7310508%	1 / 3.1514872
1.281552σ	80%	20%	1 / 5
1.644854σ	90%	10%	1 / 10
1.959964σ	95%	5%	1 / 20
2σ	95.4499736%	4.5500264%	1 / 21.977895
2.575829σ	99%	1%	1 / 100
3σ	99.7300204%	0.2699796%	1 / 370.398
3.290527σ	99.9%	0.1%	1 / 1000
3.890592σ	99.99%	0.01%	1 / 10000
4σ	99.993666%	0.006334%	1 / 15787
4.417173σ	99.999%	0.001%	1 / 100000
4.5σ	99.9993204653751%	0.0006795346249%	3.4 / 1000000 (on each side of mean)
4.891638σ	99.9999%	0.0001%	1 / 1000000
5σ	99.9999426697%	0.0000573303%	1 / 1744278
5.326724σ	99.99999%	0.00001%	1 / 10000000
5.730729σ	99.999999%	0.000001%	1 / 100000000
6σ	99.9999998027%	0.0000001973%	1 / 506797346
6.109410σ	99.9999999%	0.0000001%	1 / 1000000000
6.466951σ	99.99999999%	0.00000001%	1 / 10000000000

| 6.806502σ | 99.999999999% | 0.000000001% | 1 / 100000000000 |
| 7σ | 99.9999999997440% | 0.000000000256% | 1 / 390682215445 |

Relationship between Standard Deviation and Mean

The mean and the standard deviation of a set of data are descriptive statistics usually reported together. In a certain sense, the standard deviation is a "natural" measure of statistical dispersion if the center of the data is measured about the mean. This is because the standard deviation from the mean is smaller than from any other point. The precise statement is the following: suppose x_1, ..., x_n are real numbers and define the function:

$$\sigma(r) = \sqrt{\frac{1}{N-1}\sum_{i=1}^{N}(x_i - r)^2}.$$

Using calculus or by completing the square, it is possible to show that $\sigma(r)$ has a unique minimum at the mean:

$$r = \overline{x}.$$

Variability can also be measured by the coefficient of variation, which is the ratio of the standard deviation to the mean. It is a dimensionless number.

Standard Deviation of the Mean

Often, we want some information about the precision of the mean we obtained. We can obtain this by determining the standard deviation of the sampled mean. Assuming statistical independence of the values in the sample, the standard deviation of the mean is related to the standard deviation of the distribution by:

$$\sigma_{mean} = \frac{1}{\sqrt{N}}\sigma$$

where N is the number of observations in the sample used to estimate the mean. This can easily be proven with:

$$\text{var}(X) \equiv \sigma_X^2$$

$$\text{var}(X_1 + X_2) \equiv \text{var}(X_1) + \text{var}(X_2)$$

(Statistical Independence is assumed.)

$$\text{var}(cX_1) \equiv c^2 \text{var}(X_1)$$

hence

$$\text{var(mean)} = \text{var}\left(\frac{1}{N}\sum_{i=1}^{N} X_i\right) = \frac{1}{N^2}\text{var}\left(\sum_{i=1}^{N} X_i\right)$$

$$= \frac{1}{N^2}\sum_{i=1}^{N}\text{var}(X_i) = \frac{N}{N^2}\text{var}(X) = \frac{1}{N}\text{var}(X).$$

Resulting in:

$$\sigma_{\text{mean}} = \frac{\sigma}{\sqrt{N}}.$$

It should be emphasized that in order to estimate standard deviation of the mean it is necessary to know standard deviation of the entire population σ_{mean} beforehand. However, in most applications this parameter is unknown. For example, if series of 10 measurements of previously unknown quantity is performed in laboratory, it is possible to calculate resulting sample mean and sample standard deviation, but it is impossible to calculate standard deviation of the mean.

Rapid Calculation Methods

The following two formulas can represent a running (repeatedly updated) standard deviation. A set of two power sums s_1 and s_2 are computed over a set of N values of x, denoted as $x_1, ..., x_N$:

$$s_j = \sum_{k=1}^{N} x_k^j.$$

Given the results of these running summations, the values N, s_1, s_2 can be used at any time to compute the *current* value of the running standard deviation:

$$\sigma = \frac{\sqrt{Ns_2 - s_1^2}}{N}$$

Where N, as mentioned above, is the size of the set of values (or can also be regarded as s_0).

Similarly for sample standard deviation,

$$s = \sqrt{\frac{Ns_2 - s_1^2}{N(N-1)}}.$$

In a computer implementation, as the three s_j sums become large, we need to consider round-off error, arithmetic overflow, and arithmetic underflow. The method below calculates the running sums method with reduced rounding errors. This is a "one pass" algorithm for calculating variance of n samples without the need to store prior data during the calculation. Applying this method to a time series will result in successive values of standard deviation corresponding to n data points as n grows larger with each new sample, rather than a constant-width sliding window calculation.

For $k = 1, ..., n$:

$$A_0 = 0$$

$$A_k = A_{k-1} + \frac{x_k - A_{k-1}}{k}$$

where A is the mean value.

$$Q_0 = 0$$

$$Q_k = Q_{k-1} + \frac{k-1}{k}(x_k - A_{k-1})^2 = Q_{k-1} + (x_k - A_{k-1})(x_k - A_k)$$

Note: $Q_1 = 0$ since $k - 1 = 0$ or $x_1 = A_1$

Sample variance:

$$s_n^2 = \frac{Q_n}{n-1}$$

Population variance:

$$n_n^2 = \frac{Q_n}{n}$$

Weighted Calculation

When the values x_i are weighted with unequal weights w_i, the power sums s_0, s_1, s_2 are each computed as:

$$s_j = \sum_{k=1}^{N} w_k x_k^j.$$

And the standard deviation equations remain unchanged. Note that s_0 is now the sum of the weights and not the number of samples N.

The incremental method with reduced rounding errors can also be applied, with some additional complexity.

A running sum of weights must be computed for each k from 1 to n:

$$W_0 = 0$$

$$W_k = W_{k-1} + w_k$$

and places where $1/n$ is used above must be replaced by w_i/W_n:

$$A_0 = 0$$

$$A_k = A_{k-1} + \frac{w_k}{W_k}(x_k - A_{k-1})$$

$$Q_0 = 0$$

$$Q_k = Q_{k-1} + \frac{w_k W_{k-1}}{W_k}(x_k - A_{k-1})^2 = Q_{k-1} + w_k(x_k - A_{k-1})(x_k - A_k)$$

In the final division,

$$\sigma_n^2 = \frac{Q_n}{W_n}$$

and

$$s_n^2 = \frac{Q_n}{W_n - 1},$$

where n is the total number of elements, and n' is the number of elements with non-zero weights. The above formulas become equal to the simpler formulas given above if weights are taken as equal to one.

History

The term *standard deviation* was first used in writing by Karl Pearson in 1894, following his use of it in lectures. This was as a replacement for earlier alternative names for the same idea: for example, Gauss used *mean error*. It may be worth noting in passing that the mean error is mathematically distinct from the standard deviation.

Average Absolute Deviation

The average absolute deviation (or mean absolute deviation) of a data set is the average of the absolute deviations from a central point. It is a summary statistic of statistical dispersion or variability. In this general form, the central point can be the mean, median, mode, or the result of another measure of central tendency. Furthermore, as described in the article about averages, the deviation averaging operation may refer to the mean or the median. Thus the total number of combinations amounts to at least four types of average absolute deviation.

Measures of Dispersion

Several measures of statistical dispersion are defined in terms of the absolute deviation. The term "average absolute deviation" does not uniquely identify a measure of statistical dispersion, as there are several measures that can be used to measure absolute deviations, and there are several measures of central tendency that can be used as well. Thus, to uniquely identify the absolute deviation it is necessary to specify both the measure of deviation and the measure of central tendency. Unfortunately, the statistical literature has not yet adopted a standard notation, as both the #Mean absolute deviation around the mean and the #Median absolute deviation around the median have been denoted by their initials "MAD" in the literature, which may lead to confusion, since in general, they may have values considerably different from each other.

Mean Absolute Deviation Around a Central Point

The mean absolute deviation of a set $\{x_1, x_2, ..., x_n\}$ is

$$\frac{1}{n}\sum_{i=1}^{n}|x_i - m(X)|.$$

The choice of measure of central tendency, $m(X)$, has a marked effect on the value of the mean deviation. For example, for the data set $\{2, 2, 3, 4, 14\}$:

Measure of central tendency $m(X)$	Mean absolute deviation
Mean = 5	$\dfrac{|2-5|+|2-5|+|3-5|+|4-5|+|14-5|}{5} = 3.6$
Median = 3	$\dfrac{|2-3|+|2-3|+|3-3|+|4-3|+|14-3|}{5} = 2.8$
Mode = 2	$\dfrac{|2-2|+|2-2|+|3-2|+|4-2|+|14-2|}{5} = 3.0$

The mean absolute deviation from the median is less than or equal to the mean absolute deviation from the mean. In fact, the mean absolute deviation from the median is always less than or equal to the mean absolute deviation from any other fixed number.

The mean absolute deviation from the mean is less than or equal to the standard deviation; one way of proving this relies on Jensen's inequality.

Proof

Jensen's inequality is $\varphi\big(\mathbb{E}[X]\big) \leq \mathbb{E}\big[\varphi(X)\big]$, where φ is a convex function, this implies that:

$$\mathbb{E}(|\,x - \mu\,|)^2 \leq \mathbb{E}\big(|\,x - \mu\,|^2\big)$$

$$\mathbb{E}(|\,x - \mu\,|)^2 \leq \mathrm{Var}(x)$$

Since both sides are positive, and the square root is a monotonically increasing function in the positive domain:

$$\mathbb{E}(|\,x - \mu\,|) \leq \sqrt{\mathrm{Var}(x)}$$

For a general case of this statement

For the normal distribution, the ratio of mean absolute deviation to standard deviation is $\sqrt{2/\pi} = 0.79788456\ldots$ Thus if X is a normally distributed random variable with expected value o then:

$$w = \frac{E\,|\,X\,|}{\sqrt{E(X^2)}} = \sqrt{\frac{2}{\pi}}.$$

In other words, for a normal distribution, mean absolute deviation is about 0.8 times the standard deviation. However in-sample measurements deliver values of the ratio of mean average deviation / standard deviation for a given Gaussian sample n with the following bounds: $w_n \in [0,1]$, with a bias for small n.

Mean Absolute Deviation Around the Mean

The mean absolute deviation (MAD), also referred to as the "mean deviation" or sometimes "average absolute deviation", is the mean of the data's absolute deviations around the data's mean: the average (absolute) distance from the mean. "Average absolute deviation" can refer to either this usage, or to the general form with respect to a specified central point.

MAD has been proposed to be used in place of standard deviation since it corresponds better to real life. Because the MAD is a simpler measure of variability than the standard deviation, it can be used as pedagogical tool to help motivate the standard deviation.

This method's forecast accuracy is very closely related to the mean squared error (MSE) method which is just the average squared error of the forecasts. Although these methods are very closely related, MAD is more commonly used because it is both easier to compute (avoiding the need for squaring) and easier to understand.

Mean Absolute Deviation Around the Median

Mean absolute deviation around the median (MAD median) offers a direct measure of the scale of a random variable around its median

$$D_{med} = E \,|\, X - \text{median} \,|$$

For the normal distribution we have $D_{med} = \sigma\sqrt{2/\pi}$. Since the median minimizes the average absolute distance, we have $D_{med} \leq D_{mean}$. By using the general dispersion function Habib (2011) defined MAD about median as

$$D_{med} = E \,|\, X - \text{median} \,| = 2\,\text{Cov}(X, I_O)$$

where the indicator function is

$$\mathbf{I}_O := \begin{cases} 1 & \text{if } x > \text{median}, \\ 0 & \text{otherwise.} \end{cases}$$

This representation allows for obtaining MAD median correlation coefficients;

Median Absolute Deviation Around a Central Point

Median Absolute Deviation Around the Mean

In principle the mean could be taken as the central point for the median absolute deviation, but more often the median value is taken instead.

Median Absolute Deviation Around the Median

The median absolute deviation (also MAD) is the *median* of the absolute deviation from the *median*. It is a robust estimator of dispersion.

For the example {2, 2, 3, 4, 14}: 3 is the median, so the absolute deviations from the median are {1, 1, 0, 1, 11} (reordered as {0, 1, 1, 1, 11}) with a median of 1, in this case unaffected by the value of the outlier 14, so the median absolute deviation (also called MAD) is 1.

Maximum Absolute Deviation

The maximum absolute deviation around an arbitrary point is the maximum of the absolute deviations of a sample from that point. While not strictly a measure of central tendency, the maximum absolute deviation can be found using the formula for the average absolute deviation as above with $m(X) = \max(X)$, where $\max(X)$ is the sample maximum.

Minimization

The measures of statistical dispersion derived from absolute deviation characterize various measures of central tendency as *minimizing* dispersion: The median is the measure of central tendency most associated with the absolute deviation. Some location parameters can be compared as follows:

- L^2 norm statistics: the mean minimizes the mean squared error

- L^1 norm statistics: the median minimizes *average* absolute deviation,

- L^∞ norm statistics: the mid-range minimizes the *maximum* absolute deviation

- trimmed L^∞ norm statistics: for example, the midhinge (average of first and third quartiles) which minimizes the *median* absolute deviation of the whole distribution, also minimizes the *maximum* absolute deviation of the distribution after the top and bottom 25% have been trimmed off..

Estimation

The mean absolute deviation of a sample is a biased estimator of the mean absolute deviation of the population. In order for the absolute deviation to be an unbiased estimator, the expected value (average) of all the sample absolute deviations must equal the population absolute deviation. However, it does not. For the population 1,2,3 both the population absolute deviation about the median and the population absolute deviation about the mean are 2/3. The average of all the sample absolute deviations about the mean of size 3 that can be drawn from the population is 44/81, while the average of all the sample absolute deviations about the median is 4/9. Therefore the absolute deviation is a biased estimator. However, this argument is based on the notion of mean-unbiasedness. Each measure of location has its own form of unbiasedness. The relevant form of unbiasedness here is median unbiasedness.

Median Absolute Deviation

In statistics, the median absolute deviation (MAD) is a robust measure of the variability of a univariate sample of quantitative data. It can also refer to the population parameter that is estimated by the MAD calculated from a sample.

For a univariate data set $X_1, X_2, ..., X_n$, the MAD is defined as the median of the absolute deviations from the data's median:

$$MAD = median\left(\left| X_i - median(X) \right| \right),$$

that is, starting with the residuals (deviations) from the data's median, the MAD is the median of their absolute values.

Example

Consider the data (1, 1, 2, 2, 4, 6, 9). It has a median value of 2. The absolute deviations about 2 are (1, 1, 0, 0, 2, 4, 7) which in turn have a median value of 1 (because the sorted absolute deviations are (0, 0, 1, 1, 2, 4, 7)). So the median absolute deviation for this data is 1.

Uses

The median absolute deviation is a measure of statistical dispersion. Moreover, the MAD is a robust statistic, being more resilient to outliers in a data set than the standard deviation. In the standard deviation, the distances from the mean are squared, so large deviations are weighted more heavily, and thus outliers can heavily influence it. In the MAD, the deviations of a small number of outliers are irrelevant.

Because the MAD is a more robust estimator of scale than the sample variance or standard deviation, it works better with distributions without a mean or variance, such as the Cauchy distribution.

Relation to Standard Deviation

In order to use the MAD as a consistent estimator for the estimation of the standard deviation σ, one takes

$$\hat{\sigma} = k \cdot MAD,$$

where k is a constant scale factor, which depends on the distribution.

For normally distributed data k is taken to be:

$$k = 1/\left(\Phi^{-1}(3/4)\right) \approx 1.4826,$$

i.e., the reciprocal of the quantile function Φ^{-1} (also known as the inverse of the cumulative distribution function) for the standard normal distribution $Z = X/\sigma$. The argument 3/4 is such that $\pm MAD$ covers 50% (between 1/4 and 3/4) of the standard normal cumulative distribution function, i.e.:

$$\frac{1}{2} = P(|X - \mu| \le \text{MAD}) = P\left(\left|\frac{X - \mu}{\sigma}\right| \le \frac{\text{MAD}}{\sigma}\right) = P\left(|Z| \le \frac{\text{MAD}}{\sigma}\right).$$

Therefore, we must have that:

$$\Phi\left(\text{MAD}/\sigma\right) - \Phi\left(-\text{MAD}/\sigma\right) = 1/2..$$

Noticing that:

$$\Phi\left(-\text{MAD}/\sigma\right) = 1 - \Phi\left(\text{MAD}/\sigma\right)$$

we have that $\text{MAD}/\sigma = \Phi^{-1}(3/4) = 0.67449$ from which we obtain the scale factor

$k = 1/\Phi^{-1}(3/4) = 1.4826..$

Another way of establishing the relationship is noting that MAD equals the half-normal distribution median:

$$MAD = \sigma\sqrt{2}\text{erf}^{-1}(1/2) \approx 0.67449\sigma..$$

This form is used in, e.g., the probable error.

The Population MAD

The population MAD is defined analogously to the sample MAD, but is based on the complete distribution rather than on a sample. For a symmetric distribution with zero mean, the population MAD is the 75th percentile of the distribution.

Unlike the variance, which may be infinite or undefined, the population MAD is always a finite number. For example, the standard Cauchy distribution has undefined variance, but its MAD is 1.

The earliest known mention of the concept of the MAD occurred in 1816, in a paper by Carl Friedrich Gauss on the determination of the accuracy of numerical observations.

Interquartile Range

In descriptive statistics, the interquartile range (IQR), also called the midspread or middle fifty, or technically H-spread, is a measure of statistical dispersion, being equal to the difference between the upper and lower quartiles, IQR = $Q_3 - Q_1$. In other words, the IQR is the 1st quartile subtracted from the 3rd quartile; these quartiles can be clearly seen on a box plot on the data. It is a trimmed estimator, defined as the 25% trimmed range, and is the most significant basic robust measure of scale.

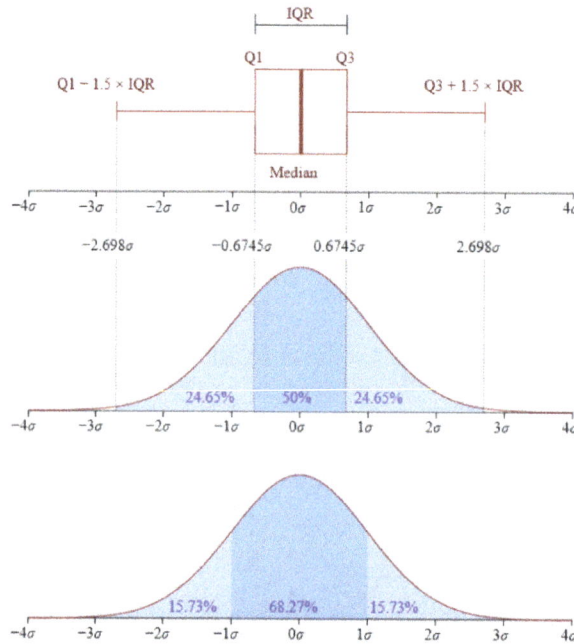

Boxplot (with an interquartile range) and a probability density function (pdf) of a Normal N(0,σ²) Population

The interquartile range (IQR) is a measure of variability, based on dividing a data set into quartiles. Quartiles divide a rank-ordered data set into four equal parts. The values that divide each part are called the first, second, and third quartiles; and they are denoted by Q1, Q2, and Q3, respectively.

Use

Unlike total range, the interquartile range has a breakdown point of 25%, and is thus often preferred to the total range.

The IQR is used to build box plots, simple graphical representations of a probability distribution.

For a symmetric distribution (where the median equals the midhinge, the average of the first and third quartiles), half the IQR equals the median absolute deviation (MAD).

The median is the corresponding measure of central tendency.

The IQR can be used to identify outliers.

The quartile deviation or semi-interquartile range is defined as half the IQR.

Examples

Data Set in a Table

i	x[i]	Quartile
1	102	
2	104	
3	105	Q_1
4	107	
5	108	
6	109	Q_2 (median)
7	110	
8	112	
9	115	Q_3
10	116	
11	118	

For the data in this table the interquartile range is IQR = 115 − 105 = 10.

Data Set in a Plain-text Box Plot

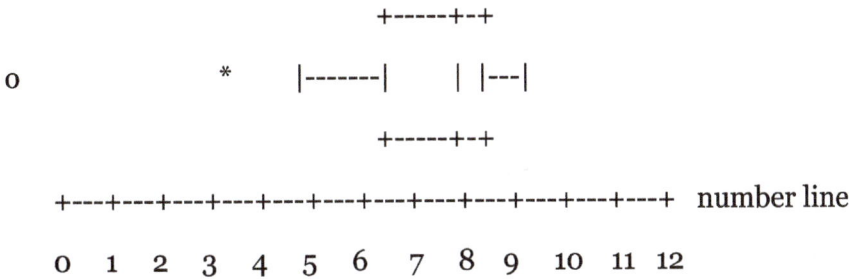

```
                        +-----+-+
  0                *    |-------|     | |---|

                        +-----+-+

      +---+---+---+---+---+---+---+---+---+---+---+---+  number line

      0   1   2   3   4   5   6   7   8   9   10  11  12
```

For the data set in this box plot:

- lower (first) quartile Q_1 = 7

- median (second quartile) Q_2 = 8.5

- upper (third) quartile Q_3 = 9

- interquartile range, IQR = $Q_3 - Q_1$ = 2

Interquartile Range of Distributions

The interquartile range of a continuous distribution can be calculated by integrating the probability density function (which yields the cumulative distribution function — any other means of calculating the CDF will also work). The lower quartile, Q_1, is a number such that integral of the PDF from -∞ to Q_1 equals 0.25, while the upper quartile, Q_3, is such a number that the integral from -∞ to Q_3 equals 0.75; in terms of the CDF, the quartiles can be defined as follows:

$$Q_1 = CDF^{-1}(0.25),$$

$$Q_3 = CDF^{-1}(0.75),$$

where CDF^{-1} is the quantile function.

The interquartile range and median of some common distributions are shown below

Distribution	Median	IQR
Normal	μ	$2\,\Phi^{-1}(0.75)\sigma \approx 1.349\sigma \approx (27/20)\sigma$
Laplace	μ	$2b\ln(2) \approx 1.386b$
Cauchy	μ	2γ

Interquartile Range test for Normality of Distribution

The IQR, mean, and standard deviation of a population P can be used in a simple test of whether or not P is normally distributed, or Gaussian. If P is normally distributed, then the standard score of the first quartile, z_1, is -0.67, and the standard score of the third quartile, z_3, is +0.67. Given *mean = X* and *standard deviation* = σ for P, if P is normally distributed, the first quartile

$$Q_1 = (\sigma z_1) + X$$

and the third quartile

$$Q_3 = (\sigma z_3) + X$$

If the actual values of the first or third quartiles differ substantially from the calculated values, P is not normally distributed.

Interquartile Range and Outliers

The interquartile range is often used to find outliers in data. Outliers are observations that fall below Q1 - 1.5(IQR) or above Q3 + 1.5(IQR). In a boxplot, the highest and lowest occurring value within this limit are drawn as bar of the *whiskers*, and the outliers as individual points.

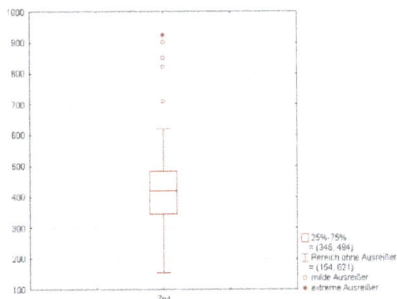

Box-and-whisker plot with four close and one far away extreme values, defined as outliers above Q3 + 1.5(IQR) and Q3 + 3(IQR), respectively.

Mean Absolute Difference

The mean absolute difference is a measure of statistical dispersion equal to the average absolute difference of two independent values drawn from a probability distribution. A related statistic is the relative mean absolute difference, which is the mean absolute difference divided by the arithmetic mean, and equal to twice the Gini coefficient. The mean absolute difference is also known as the absolute mean difference and the Gini mean absolute difference. The mean absolute difference is sometimes denoted by Δ or as MD.

Definition

The mean absolute difference or the MAD is defined as the "average" or "mean", formally the expected value, of the absolute difference of two random variables X and Y independently and identically distributed with the same (unknown) distribution henceforth called Q.

$$\mathrm{MD} := E[|X - Y|].$$

Calculation

Specifically, in the discrete case,

- For a random sample of size n of a population distributed according to Q, the (empirical) mean absolute difference of the sequence of sample values y_i, $i = 1$ to n can be calculated as the arithmetic mean of the absolute value of all possible differences:

$$\mathrm{MD} = \frac{1}{n^2} \sum_{i=1}^{n} \sum_{j=1}^{n} |y_i - y_j|.$$

- if Q has a discrete probability function $f(y)$, where y_i, $i = 1$ to n, are the values with nonzero probabilities:

$$MD = \sum_{i=1}^{n} \sum_{j=1}^{n} f(y_i)f(y_j) \, | \, y_i - y_j \, |.$$

In the continuous case,

- if Q has a probability density function $f(x)$:

$$MD = \int_{-\infty}^{\infty} \int_{-\infty}^{\infty} f(x)f(y) \, | \, x - y \, | \, dx \, dy.$$

- if Q has a cumulative distribution function $F(x)$ with quantile function $Q(F)$, then, since $f(x)=dF(x)/dx$ and $Q(F(x))=x$, it follows that:

$$MD = \int_{0}^{1} \int_{0}^{1} | \, Q(F_1) - Q(F_2) \, | \, dF_1 \, dF_2.$$

Relative Mean Absolute Difference

When the probability distribution has a finite and nonzero arithmetic mean, the relative mean absolute difference, sometimes denoted by Δ or RMD, is defined by

$$RMD = \frac{MD}{\text{arithmetic mean}}.$$

The relative mean absolute difference quantifies the mean absolute difference in comparison to the size of the mean and is a dimensionless quantity. The relative mean absolute difference is equal to twice the Gini coefficient which is defined in terms of the Lorenz curve. This relationship gives complementary perspectives to both the relative mean absolute difference and the Gini coefficient, including alternative ways of calculating their values.

Properties

The mean absolute difference is invariant to translations and negation, and varies proportionally to positive scaling. That is to say, if X is a random variable and c is a constant:

- $MD(X + c) = MD(X)$,

- $MD(-X) = MD(X)$, and

- $MD(c\,X) = |c|\,MD(X)$.

The relative mean absolute difference is invariant to positive scaling, commutes with negation, and varies under translation in proportion to the ratio of the original and translated arithmetic means. That is to say, if X is a random variable and c is a constant:

- $\text{RMD}(X + c) = \text{RMD}(X) \cdot \text{mean}(X)/(\text{mean}(X) + c) = \text{RMD}(X) / (1 + c / \text{mean}(X))$ for $c \neq -\text{mean}(X)$,

- $\text{RMD}(-X) = -\text{RMD}(X)$, and

- $\text{RMD}(c\,X) = \text{RMD}(X)$ for $c > 0$.

If a random variable has a positive mean, then its relative mean absolute difference will always be greater than or equal to zero. If, additionally, the random variable can only take on values that are greater than or equal to zero, then its relative mean absolute difference will be less than 2.

Compared to Standard Deviation

The mean absolute difference is twice the L-scale (the second L-moment), while the standard deviation is the square root of the variance about the mean (the second conventional central moment). The differences between L-moments and conventional moments are first seen in comparing the mean absolute difference and the standard deviation (the first L-moment and first conventional moment are both the mean).

Both the standard deviation and the mean absolute difference measure dispersion—how spread out are the values of a population or the probabilities of a distribution. The mean absolute difference is not defined in terms of a specific measure of central tendency, whereas the standard deviation is defined in terms of the deviation from the arithmetic mean. Because the standard deviation squares its differences, it tends to give more weight to larger differences and less weight to smaller differences compared to the mean absolute difference. When the arithmetic mean is finite, the mean absolute difference will also be finite, even when the standard deviation is infinite.

The recently introduced distance standard deviation plays similar role to the mean absolute difference but the distance standard deviation works with centered distances.

Sample Estimators

For a random sample S from a random variable X, consisting of n values y_i, the statistic

$$\text{MD}(S) = \frac{\sum_{i=1}^{n} \sum_{j=1}^{n} |y_i - y_j|}{n(n-1)}$$

is a consistent and unbiased estimator of MD(X). The statistic:

$$\text{RMD}(S) = \frac{\sum_{i=1}^{n}\sum_{j=1}^{n}|y_i - y_j|}{(n-1)\sum_{i=1}^{n}y_i}$$

is a consistent estimator of RMD(X), but is not, in general, unbiased.

Confidence intervals for RMD(X) can be calculated using bootstrap sampling techniques.

There does not exist, in general, an unbiased estimator for RMD(X), in part because of the difficulty of finding an unbiased estimation for multiplying by the inverse of the mean. For example, even where the sample is known to be taken from a random variable $X(p)$ for an unknown p, and $X(p) - 1$ has the Bernoulli distribution, so that $\Pr(X(p) = 1) = 1 - p$ and $\Pr(X(p) = 2) = p$, then

$$\text{RMD}(X(p)) = 2p(1 - p)/(1 + p).$$

But the expected value of any estimator $R(S)$ of RMD($X(p)$) will be of the form:

$$E(R(S)) = \sum_{i=0}^{n}p^i(1-p)^{n-i}r_i,$$

where the r_i are constants. So $E(R(S))$ can never equal RMD($X(p)$) for all p between 0 and 1.

Examples

Examples of mean absolute difference and relative mean absolute difference					
Distri-bution	Param-eters	Mean	Standard deviation	Mean absolute difference	Relative mean absolute difference
Con-tinuous uniform	a = 0 ; b = 1	1 / 2 = 0.5	$\frac{1}{\sqrt{12}} \approx 0.2887$	1 / 3 ≈ 0.3333	2 / 3 ≈ 0.6667
Normal	$\mu = 1$; $\sigma = 1$	1	1	$\frac{2}{\sqrt{\pi}} \approx 1.1284$	$\frac{2}{\sqrt{\pi}} \approx 1.1284$
Exponential	$\lambda = 1$	1	1	1	1
Pareto	$k > 1$; $x_m = 1$	$\frac{k}{k-1}$	$\frac{1}{k-1}\sqrt{\frac{k}{k-2}}$ (for $k > 2$)	$\frac{2k}{(k-1)(2k-1)}$	$\frac{2}{2k-1}$

Gamma	$k \, ; \theta$	$k\theta$	$\sqrt{k}\,\theta$	$k\theta(4I_{0.5}(k+1,k)-2)$ [†]	$4I_{0.5}(k+1,k)-2$ [†]
Gamma	$k=1;$ $\theta=1$	1	1	1	1
Gamma	$k=2;\theta$ $=1$	2	$\sqrt{2}\approx1.4142$	$3/2=1.5$	$3/4=0.75$
Gamma	$k=3;\theta$ $=1$	3	$\sqrt{3}\approx1.7321$	$15/8=1.875$	$5/8=0.625$
Gamma	$k=4;\theta$ $=1$	4	2	$35/16=2.1875$	$35/64=0.546875$
Bernoulli	$0\le p\le1$	p	$\sqrt{p(1-p)}$	$2p(1-p)$	$2(1-p)$ for $p>0$
Student's t, 2 d.f.	$\nu=2$	0	∞	$\pi/\sqrt{2}=2.2214$	undefined

[†] $I_z(x,y)$ is the regularized incomplete Beta function

Range (Statistics)

In arithmetic, the range of a set of data is the difference between the largest and smallest values.

However, in descriptive statistics, this concept of range has a more complex meaning. The range is the size of the smallest interval which contains all the data and provides an indication of statistical dispersion. It is measured in the same units as the data. Since it only depends on two of the observations, it is most useful in representing the dispersion of small data sets.

Independent Identically Distributed Continuous Random Variables

For n independent and identically distributed continuous random variables X_1, X_2, ..., X_n with cumulative distribution function $G(x)$ and probability density function $g(x)$ the of range the X_i is the range of a sample of size n from a population with distribution function $G(x)$.

Distribution

The range has cumulative distribution function

$$F(t) = n\int_{-\infty}^{\infty} g(x)[G(x+t)-G(x)]^{n-1}\,\mathrm{d}x.$$

Gumbel notes that the "beauty of this formula is completely marred by the facts that, in general, we cannot express $G(x + t)$ by $G(x)$, and that the numerical integration is lengthy and tiresome."

If the distribution of each X_i is limited to the right (or left) then the asymptotic distribution of the range is equal to the asymptotic distribution of the largest (smallest) value. For more general distributions the asymptotic distribution can be expressed as a Bessel function.

Moments

The mean range is given by

$$n\int_0^1 x(G)[G^{n-1} - (1-G)^{n-1}]dG$$

where $x(G)$ is the inverse function. In the case where each of the X_i has a standard normal distribution, the mean range is given by

$$\int_{-\infty}^{\infty} (1 - (1 - \Phi(x))^n - \Phi(x)^n)dx.$$

Independent Nonidentically Distributed Continuous Random Variables

For n nonidentically distributed independent continuous random variables $X_1, X_2, ..., X_n$ with cumulative distribution functions $G_1(x), G_2(x), ..., G_n(x)$ and probability density functions $g_1(x), g_2(x), ..., g_n(x)$, the range has cumulative distribution function

$$F(t) = \sum_{i=1}^{n} \int_{-\infty}^{\infty} g_i(x) \prod_{j=1, j \neq i}^{n} [G_j(x+t) - G_j(x)] dx$$

Independent Identically Distributed Discrete Random Variables

For n independent and identically distributed discrete random variables $X_1, X_2, ..., X_n$ with cumulative distribution function $G(x)$ and probability mass function $g(x)$ the range of the X_i is the range of a sample of size n from a population with distribution function $G(x)$. We can assume without loss of generality that the support of each X_i is $\{1,2,3,...,N\}$ where N is a positive integer or infinity.

Example

If we suppose that $g(x) = 1/N$, the discrete uniform distribution for all x, then we find

$$f(t) = \begin{cases} \dfrac{1}{N^{n-1}} & t = 0 \\ [4pt]\displaystyle\sum_{x=1}^{N-t}\left([\frac{t+1}{N}]^n - 2[\frac{t}{N}]^n + [\frac{t-1}{N}]^n\right) & t = 1,2,3...,N-1. \end{cases}$$

Related Quantities

The range is a simple function of the sample maximum and minimum and these are specific examples of order statistics. In particular, the range is a linear function of order statistics, which brings it into the scope of L-estimation.

References

- Johnson, Richard; Wichern, Dean (2001). Applied Multivariate Statistical Analysis. Prentice Hall. p. 76. ISBN 0-13-187715-1

- Dodge, Yadolah (2003). The Oxford Dictionary of Statistical Terms. Oxford University Press. ISBN 0-19-920613-9.

- Hoaglin, David C.; Frederick Mosteller; John W. Tukey (1983). Understanding Robust and Exploratory Data Analysis. John Wiley & Sons. pp. 404–414. ISBN 0-471-09777-2.

- Russell, Roberta S.; Bernard W. Taylor III. (2006). Operations Management. John Wiley & Sons. pp. 497–498. ISBN 0-471-69209-3.

- Venables, W.N.; B.D. Ripley (1999). Modern Applied Statistics with S-PLUS. Springer. p. 128. ISBN 0-387-98825-4.

- Upton, Graham; Cook, Ian (1996). Understanding Statistics. Oxford University Press. p. 55. ISBN 0-19-914391-9.

- Zwillinger, D., Kokoska, S. (2000) CRC Standard Probability and Statistics Tables and Formulae, CRC Press. ISBN 1-58488-059-7 page 18.

- "CERN experiments observe particle consistent with long-sought Higgs boson | CERN press office". Press.web.cern.ch. 2012-07-04. Retrieved 2015-05-30.

- Kader, Gary (March 1999). "Means and MADS". Mathematics Teaching in the Middle School. 4 (6): 398–403. Retrieved 20 February 2013.

Applications of Statistics

The acquiring of information from a given population is census whereas actuarial science assesses risk in the sectors of finance and insurance. The applications of statistics discussed in this chapter are census, actuarial science, demography, environmental studies etc. The aspects elucidated in this chapter are of vital importance, and provide a better understanding of statistics.

Census

Census taker visits a Romani family living in a caravan, Netherlands 1925

A census is the procedure of systematically acquiring and recording information about the members of a given population. It is a regularly occurring and official count of a particular population. The term is used mostly in connection with national population and housing censuses; other common censuses include agriculture, business, and traffic censuses. The United Nations defines the essential features of population and housing censuses as "individual enumeration, universality within a defined territory, simultaneity and defined periodicity", and recommends that population censuses be taken at least every 10 years. United Nations recommendations also cover census topics to be collected, official definitions, classifications and other useful information to co-ordinate international practice.

The word is of Latin origin: during the Roman Republic, the census was a list that kept track of all adult males fit for military service. The modern census is essential to international comparisons of any kind of statistics, and censuses collect data on many attributes of a population, not just how many people there are but now census takes its place within a system of surveys where it typically began as the only national demographic data collection. Although population estimates remain an important function of a census, including exactly the geographic distribution of the population, statistics can be produced about combinations of attributes e.g. education by age and sex in different regions. Current administrative data systems allow for other approaches to enumeration with the same level of detail but raise concerns about privacy and the possibility of biasing estimates.

The United Nations Population Fund (UNFPA) explains that, "A traditional population and housing census requires mapping an entire country, figuring out what technologies should be employed, mobilizing and training legions of enumerators, conducting a major public campaign, canvassing all households, collecting individual information, compiling hundreds of thousands – or millions – of completed questionnaires, monitoring procedures and results, and analyzing and disseminating the data."

A census can be contrasted with sampling in which information is obtained only from a subset of a population, typically main population estimates are updated by such intercensal estimates. Modern census data are commonly used for research, business marketing, and planning, and as a baseline for designing sample surveys by providing a sampling frame such as an address register. Census counts are necessary to adjust samples to be representative of a population by weighting them as is common in opinion polling. Similarly, stratification requires knowledge of the relative sizes of different population strata which can be derived from census enumerations. In some countries, the census provides the official counts used to apportion the number of elected representatives to regions (sometimes controversially – e.g., *Utah v. Evans*). In many cases, a carefully chosen random sample can provide more accurate information than attempts to get a population census.

Sampling

A census is often construed as the opposite of a sample as its intent is to count everyone in a population rather than a fraction. However, population censuses relies on a sampling frame to count the population. This is the only way to be sure that everyone has been included as otherwise those not responding would not be followed up on and individuals could be missed. The fundamental premise of a census is that the population is not known and a new estimate is to be made by the analysis of primary data. The use of a sampling frame is counterintuitive as it suggests that the population size is already known. However, a census is also used to collect attribute data on the individuals in the nation. This process of sampling marks the difference between historical census, which was a house to house process or the product of an imperial decree, and the modern statistical project. The sampling frame used by census is almost always an address register. Thus it is not known if there is anyone resident or how many people there are in each household. Depending on the mode of enumeration, a form is sent to the householder, an enumerator calls, or administrative records for the dwelling are accessed. As a preliminary to the dispatch of forms, census workers will check any address problems on the ground. While it may seem straightforward to use the postal service file for this purpose, this can be out of date and some dwellings may contain a number of independent households. A particular problem is what are termed 'communal establishments' which category includes student residences, religious orders, homes for the elderly, people in prisons etc. As these are not easily enumerated by a single householder, they are often treated differently and visited by special teams of census workers to ensure they are classified appropriately.

Residence Definitions

Individuals are normally counted within households and information is typically collected about the household structure and the housing. For this reason international documents refer to censuses of population and housing. Normally the census response is made by a household, indicating details of individuals resident there. An important aspect of census enumerations is determining which individuals can be counted from which cannot be counted. Broadly, three definitions can be used: *de facto* residence; *de jure* residence; and, permanent residence. This is important to consider individuals who have multiple or temporary addresses. Every person should be identified uniquely as resident in one place but where they happen to be on Census Day, their *de facto* residence, may not be the best place to count them. Where an individual uses services may be more useful and this is at their usual, or *de jure*, residence. An individual may be represented at a permanent address, perhaps a family home for students or long term migrants. It is necessary to have a precise definition of residence to decide whether visitors to a country should be included in the population count. This is becoming more important as students travel abroad for education for a period of several years. Other groups causing problems of enumeration are new born babies, refugees, people away on holiday, people moving home around census day, and people without a fixed address. People having second homes because of working in another part of the country or retaining a holiday cottage are difficult to fix at a particular address sometimes causing double counting or houses being mistakenly identified as vacant. Another problem is where people use a different address at different times e.g. students living at their place of education in term time but returning to a family home during vacations or children whose parents have separated who effectively have two family homes. Census enumeration has always been based on finding people where they live as there is no systematic alternative - any list you could use to find people is derived from census activities in the first place. Recent UN guidelines provide recommendation on enumerating such complex households.

Enumeration Strategies

Historical censuses used crude enumeration assuming absolute accuracy. Modern approaches take into account the problems of overcount and undercount, and the coherence of census enumerations with other official sources of data. This reflects a realist approach to measurement, acknowledging that under any definition of residence there is a true value of the population but this can never be measured with complete accuracy. An important aspect of the census process is to evaluate the quality of the data.

Many countries use a post-enumeration survey to adjust the raw census counts. This works in a similar manner to capture-recapture estimation for animal populations. In census circles this method is called dual system enumeration (DSE). A sample of households are visited by interviewers who record the details of the household as at census day. These data are then matched to census records and the number of people missed can be estimated by considering the number missed in the census or survey but counted in the other. This way counts can be adjusted for non-response varying between different demographic groups. An explanation using a fishing analogy can be found in "Trout, Catfish and Roach..." which won an award from the Royal Statistical Society for excellence in official statistics in 2011.

Triple system enumeration has been proposed as an improvement as it would allow evaluation of the statistical dependence of pairs of sources. However, as the matching process is the most difficult aspect of census estimation this has never been implemented for a national enumeration. It would also be difficult to identify three different sources that were sufficiently different to make the triple system effort worthwhile. The DSE approach has another weakness in that it assumes there is no person counted twice (over count). In *de facto* residence definitions this would not be a problem but in *de jure* definitions individuals risk being recorded on more than one form leading to double counting. A particular problem here are students who often have a term time and family address.

Several countries have used a system which is known as short form/long form. This is a sampling strategy which randomly chooses a proportion of people to send a more detailed questionnaire to (the long form). Everyone receives the short form questions. Thereby more data are collected but not imposing a burden on the whole population. This also reduces the burden on the statistical office. Indeed, in the UK all residents were required to fill in the whole form but only a 10% sample were coded and analysed in detail, until 2001. New technology means that all data are now scanned and processed. Recently there has been controversy in Canada about the cessation of the long form with the head, Munir Sheikh resigning.

The use of alternative enumeration strategies is increasing but these are not so simple as many people assume and only occur in developed countries. The Netherlands has been most advanced in adopting a census using administrative data. This allows a simulated census to be conducted by linking several different administrative databases at an agreed time. Data can be matched and an overall enumeration established accounting for where the different sources are discrepant. A validation survey is still conducted in a similar way to the post enumeration survey employed in a traditional census. Other countries which have a population register use this as a basis for all the census statistics needed by users. This is most common amongst Nordic countries but requires a large number of different registers to be combined including population, housing, employment and education. These registers are then combined and brought up to the standard of a statistical register by comparing the data in different sources and ensuring the quality is sufficient for official statistics to be produced. A recent innovation is the French instigation of a rolling census programme with different regions enumerated each year such that the whole country is completely enumerated every 5 to 10 years. In Europe, in connection with the 2010 census round, a large number of countries adopted alternative census methodologies, often based on the combination of data from registers, surveys and other sources.

Technology

Censuses have evolved in their use of technology with the latest censuses, the 2010 round, using many new types of computing. In Brazil, handheld devices were used by enumerators to locate residences on the ground. In many countries, census returns could be made via the Internet as well as in paper form. DSE is facilitated by computer matching techniques which can be automated, such as propensity score matching. In the UK, all census formats are scanned and stored electronically before being destroyed, replacing the need for physical archives. The record linking to perform an administrative census would not be possible without large databases being stored on computer systems.

New technology is not without problems in its introduction. The US census had intended to use the handheld computers but cost escalated and this was abandoned, with the contract being sold to Brazil. Online response is a good idea but one of the functions of census is to make sure everyone is counted accurately. A system which allowed people to enter their address without verification would be open to abuse. Therefore, households have to be verified on the ground, typically by an enumerator visit or post out. Paper forms are still necessary for those without access to Internet connections. It is also plausible that the hidden nature of an administrative census means that users are not engaged with the importance of contributing their data to official statistics.

Alternatively, population estimations may be carried out remotely with GIS and remote sensing technologies.

Census and Development

According to UNFPA, "The information generated by a population and housing census – numbers of people, their distribution, their living conditions and other key data – is critical for development." This is because this type of data is essential for policymakers so that they know where to invest. Unfortunately, many countries have outdated or inaccurate data about their populations and therefore, without accurate data are unable to address the needs of their population.

UNFPA stated that,

"The unique advantage of the census is that it represents the entire statistical universe, down to the smallest geographical units, of a country or region. Planners need this information for all kinds of development work, including: assessing demographic trends; analysing socio-economic conditions; designing evidence-based poverty-reduction strategies; monitoring and evaluating the effectiveness of policies; and tracking progress toward national and internationally agreed development goals."

In addition to making policymakers aware about population issues, it is also an important tool for identifying forms of social, demographic or economic exclusions, such as inequalities relating to race, ethics and religion as well as disadvantaged groups such as those with disabilities and the poor.

An accurate census can empower local communities by providing them with the necessary information to participate in local decision-making and ensuring they are represented.

Uses of Census Data

In the nineteenth century, the first censuses collected paper enumerations that had to be collated by hand so the statistical uses were very basic. The government owned the data and were able to publish statistics themselves on the state of the nation. Uses were to measure changes in the population and apportion representation. Population estimates could be compared to those of other countries.

By the beginning of the twentieth century, censuses were recording households and some indications of their employment. In some countries, census archives are released for public examination after many decades, allowing genealogists to track the ancestry of interested people. Archives provide a substantial historical record which may challenge established notions of tradition. It is also possible to understand the societal history through job titles and arrangements for the destitute and sick.

Census Data and Research

As governments assumed responsibility for schooling and welfare, large government research departments made extensive use of census data. Actuarial estimates could be made to project populations and plan for provision in local government and regions. It was also possible for central government to allocate funding on the basis of census data. Even into the mid twentieth century, census data was only directly accessible to large government departments. However, computers meant that tabulations could be used directly by university researchers, large businesses and local government offices. They could use the detail of the data to answer new questions and add to local and specialist knowledge.

Now, census data are published in a wide variety of formats to be accessible to business, all levels of governance, media, students and teachers, charities and any citizen who is interested; researchers in particular have an interest in the role of *Census Field Officers* (CFO) and their assistants. Data can be represented visually or analysed in complex statistical models, to show the difference between certain areas, or to understand the association between different personal characteristics. Census data offer a unique insight into small areas and small demographic groups which sample data would be unable to capture with precision.

Privacy

Although the census provides a useful way of obtaining statistical information about a population, such information can sometimes lead to abuses, political or otherwise, made possible by the linking of individuals' identities to anonymous census data. This consideration is particularly important when individuals' census responses are made available in microdata form, but even aggregate-level data can result in privacy breaches when dealing with small areas and/or rare subpopulations.

For instance, when reporting data from a large city, it might be appropriate to give the average income for black males aged between 50 and 60. However, doing this for a town that only has two black males in this age group would be a breach of privacy because either of those persons, knowing his own income and the reported average, could determine the other man's income.

Typically, census data are processed to obscure such individual information. Some agencies do this by intentionally introducing small statistical errors to prevent the identification of individuals in marginal populations; others swap variables for similar respondents. Whatever measures have been taken to reduce the privacy risk in census data, new technology in the form of better electronic analysis of data poses increasing challenges to the protection of sensitive individual information. This is known as statistical disclosure control.

Another possibility is to present survey results by means of statistical models in the form of a multivariate distribution mixture. The statistical information in the form of conditional distributions (histograms) can be derived interactively from the estimated mixture model without any further access to the original database. As the final product does not contain any protected microdata, the model based interactive software can be distributed without any confidentiality concerns.

Another method is simply to release no data at all, except very large scale data directly to the central government. Different release strategies between government have led to an international project (IPUMS) to co-ordinate access to microdata and corresponding metadata. Such projects also promote standardising metadata by projects such as SDMX so that best use can be made of the minimal data available.

Historical Examples

Egypt

Censuses in Egypt are said to have been taken during the early Pharaonic period in 3340 BCE and in 3056 BCE.

Ancient Greece

There are several accounts of ancient Greek and Mesopotamian city states carrying out censuses. The question of which is first is clouded by very different approaches: counting only men, counting a pile of rocks etc. but such censuses took place circa 1600 BCE and earlier.

Ancient Israel

Censuses are mentioned in the Bible. God commands a per capita tax to be paid with the census in Exodus 30:11-16 for the upkeep of the Tabernacle. The Book of Numbers is named after the counting of the Israelite population (in Numbers 1-4) according to the house of the Fathers after the exodus from Egypt. A second census was taken while the Israelite were camped in the plains of Moab, in Numbers 26.

King David performed a census that produced disastrous results (in 2 Samuel 24 and 1 Chronicles 21). His son, King Solomon, had all of the foreigners in Israel counted in 2 Chronicles 2:17.

When the Romans took over Judea in 6CE, the legate Publius Sulpicius Quirinius organised a census for tax purposes. The Gospel of Luke links the birth of Jesus to this event. Luke 2.

China

One of the world's earliest preserved censuses was held in China in 2 CE during the Han Dynasty, and is still considered by scholars to be quite accurate. Another census was held in 144 CE.

India

The oldest recorded census in India is thought to have occurred around 300 BCE during the reign of the Emperor Chandragupta Maurya under the leadership of Kautilya or Chanakya.

Rome

The word "census" originated in ancient Rome from the Latin word *censere* ("to estimate"). The census played a crucial role in the administration of the Roman Empire, as it was used to determine taxes. With few interruptions, it was usually carried out every five years. It provided a register of citizens and their property from which their duties and privileges could be listed. It is said to have been instituted by the Roman king Servius Tullius in the 6th century BCE, at which time the number of arms-bearing citizens was supposedly counted at around 80,000.

Rashidun and Umayyad Caliphates

In the Middle Ages, the Caliphate began conducting regular censuses soon after its formation, beginning with the one ordered by the second Rashidun caliph, Umar.

Medieval Europe

The Domesday Book was undertaken in 1086 CE by William I of England so that he could properly tax the land he had recently conquered in medieval Europe. In 1183 CE, a census was taken of the crusader Kingdom of Jerusalem, to ascertain the number of men and amount of money that could possibly be raised against an invasion by Saladin, sultan of Egypt and Syria.

Inca Empire

In the 15th century, the Inca Empire had a unique way to record census information. The Incas did not have any written language but recorded information collected during censuses and other numeric information as well as non-numeric data on quipus, strings from llama or alpaca hair or cotton cords with numeric and other values encoded by knots in a base-10 positional system.

Spanish Empire

On May 25, 1577, King Philip II of Spain ordered by royal cédula the preparation of a general description of Spain's holdings in the Indies. Instructions and a questionnaire, issued in 1577 by the Office of the Cronista Mayor, were distributed to local officials in the Viceroyalties of New Spain and Peru to direct the gathering of information. The questionnaire, composed of fifty items, was designed to elicit basic information about the nature of the land and the life of its peoples. The replies, known as "relaciones geográficas," were written between 1579 and 1585 and were returned to the Cronista Mayor in Spain by the Council of the Indies.

World Population Estimates

The earliest estimate of the world population was made by Giovannu Battista Riccioli in 1661; the next by Johann Peter Süssmilch in 1741, revised in 1762; the third by Karl Friedrich Wilhelm Dieterici in 1859.

In 1931 Walter Willcox published a table in his book, *International Migrations: Vol-*

ume II Interpretations, that estimated the 1929 world population to be roughly 1.8 billion.

TABLE 3.

PRESENT POPULATION OF THE EARTH AND THE CONTINENTS
(In Millions)

Continent	According to	
	International Statistical Institute 1929	League of Nations 1929
Asia	954	918
Europe	478	520
North America	162	161
Africa	140	146
South America	77	79
Australasia	9	9
Total	1,820	1,833

League of Nations and International Statistical Institute estimates of the world population in 1929

Modern Implementation

Nigerian leaders cannot put a number on the amount of Nigerian women and girls that have gone missing. Nigeria has never had a credible, successful census. —Olúfémi Táíwò, professor of Africana studies at Cornell University

Actuarial Science

2003 US mortality (life) table, Table 1, Page 1

Actuarial science is the discipline that applies mathematical and statistical methods to assess risk in insurance, finance and other industries and professions. Actuaries are professionals who are qualified in this field through intense education and experience. In many countries, actuaries must demonstrate their competence by passing a series of thorough professional examinations.

Actuarial science includes a number of interrelated subjects, including mathematics, probability theory, statistics, finance, economics, and computer science. Historically, actuarial science used deterministic models in the construction of tables and premiums. The science has gone through revolutionary changes during the last 30 years due to the proliferation of high speed computers and the union of stochastic actuarial models with modern financial theory (Frees 1990).

Many universities have undergraduate and graduate degree programs in actuarial science. In 2010, a study published by job search website CareerCast ranked actuary as the #1 job in the United States (Needleman 2010). The study used five key criteria to rank jobs: environment, income, employment outlook, physical demands, and stress. A similar study by U.S. News & World Report in 2006 included actuaries among the 25 Best Professions that it expects will be in great demand in the future (Nemko 2006).

Life Insurance, Pensions and Healthcare

Actuarial science became a formal mathematical discipline in the late 17th century with the increased demand for long-term insurance coverage such as Burial, Life insurance, and Annuities. These long term coverage required that money be set aside to pay future benefits, such as annuity and death benefits many years into the future. This requires estimating future contingent events, such as the rates of mortality by age, as well as the development of mathematical techniques for discounting the value of funds set aside and invested. This led to the development of an important actuarial concept, referred to as the Present value of a future sum. Certain aspects of the actuarial methods for discounting pension funds have come under criticism from modern financial economics.

- In traditional life insurance, actuarial science focuses on the analysis of mortality, the production of life tables, and the application of compound interest to produce life insurance, annuities and endowment policies. Contemporary life insurance programs have been extended to include credit and mortgage insurance, key man insurance for small businesses, long term care insurance and health savings accounts (Hsiao 2001).

- In health insurance, including insurance provided directly by employers, and social insurance, actuarial science focuses on the analysis of rates of disability, morbidity, mortality, fertility and other contingencies. The effects of consumer choice and the geographical distribution of the utilization of medical services and procedures, and the utilization of drugs and therapies, is also of great importance. These factors underlay the development of the Resource-Base Relative Value Scale (RBRVS) at Harvard in a multi-disciplined study (Hsiao 2004). Actuarial science also aids in the design of benefit structures, reimbursement standards, and the effects of proposed government standards on the cost of healthcare (CHBRP 2004).

- In the pension industry, actuarial methods are used to measure the costs of alternative strategies with regard to the design, funding, accounting, administration, and maintenance or redesign of pension plans. The strategies are greatly influenced by short-term and long-term bond rates, the funded status of the pension and benefit arrangements, collective bargaining; the employer's old, new and foreign competitors; the changing demographics of the workforce; changes in the internal revenue code; changes in the attitude of the internal revenue service regarding the calculation of surpluses; and equally importantly, both the short and long term financial and economic trends. It is common with mergers and acquisitions that several pension plans have to be combined or at least administered on an equitable basis. When benefit changes occur, old and new benefit plans have to be blended, satisfying new social demands and various government discrimination test calculations, and providing employees and retirees with understandable choices and transition paths. Benefit plans liabilities have to be properly valued, reflecting both earned benefits for past service, and the benefits for future service. Finally, funding schemes have to be developed that are manageable and satisfy the standards board or regulators of the appropriate country, such as the Financial Accounting Standards Board in the United States.

- In social welfare programs, the Office of the Chief Actuary (OCACT), Social Security Administration plans and directs a program of actuarial estimates and analyses relating to SSA-administered retirement, survivors and disability insurance programs and to proposed changes in those programs. It evaluates operations of the Federal Old-Age and Survivors Insurance Trust Fund and the Federal Disability Insurance Trust Fund, conducts studies of program financing, performs actuarial and demographic research on social insurance and related program issues involving mortality, morbidity, utilization, retirement, disability, survivorship, marriage, unemployment, poverty, old age, families with children, etc., and projects future workloads. In addition, the Office is charged with conducting cost analyses relating to the Supplemental Security Income (SSI) program, a general-revenue financed, means-tested program for low-income aged, blind and disabled people. The Office provides technical and consultative services to the Commissioner, to the Board of Trustees of the Social Security Trust Funds, and its staff appears before Congressional Committees to provide expert testimony on the actuarial aspects of Social Security issues.

Actuarial Science Applied to other forms of Insurance

Actuarial science is also applied to Property, Casualty, Liability, and General insurance. In these forms of insurance, coverage is generally provided on a renewable period, (such as a yearly). Coverage can be cancelled at the end of the period by either party.

Property and casualty insurance companies tend to specialize because of the complexity and diversity of risks. One division is to organize around personal and commercial lines of insurance. Personal lines of insurance are for individuals and include fire, auto, homeowners, theft and umbrella coverages. Commercial lines address the insurance needs of businesses and include property, business continuation, product liability, fleet/commercial vehicle, workers compensation, fidelity & surety, and D&O insurance. The insurance industry also provides coverage for exposures such as catastrophe, weather-related risks, earthquakes, patent infringement and other forms of corporate espionage, terrorism, and "one-of-a-kind" (e.g., satellite launch). Actuarial science provides data collection, measurement, estimating, forecasting, and valuation tools to provide financial and underwriting data for management to assess marketing opportunities and the nature of the risks. Actuarial science often helps to assess the overall risk from catastrophic events in relation to its underwriting capacity or surplus.

In the reinsurance fields, actuarial science can be used to design and price reinsurance and retrocession arrangements, and to establish reserve funds for known claims and future claims and catastrophes.

Development

Pre-formalization

Elementary mutual aid agreements and pensions arose in antiquity (Thucydides). Early in the Roman empire, associations were formed to meet the expenses of burial, cremation, and monuments—precursors to burial insurance and friendly societies. A small sum was paid into a communal fund on a weekly basis, and upon the death of a member, the fund would cover the expenses of rites and burial. These societies sometimes sold shares in the building of columbāria, or burial vaults, owned by the fund—the precursor to mutual insurance companies (Johnston 1932, §475–§476). Other early examples of mutual surety and assurance pacts can be traced back to various forms of fellowship within the Saxon clans of England and their Germanic forbears, and to Celtic society (Loan 1992). However, many of these earlier forms of surety and aid would often fail due to lack of understanding and knowledge (Faculty and Institute of Actuaries 2004).

Initial Development

The 17th century was a period of advances in mathematics in Germany, France and England. At the same time there was a rapidly growing desire and need to place the valuation of personal risk on a more scientific basis. Independently of each other, compound interest was studied and probability theory emerged as a well-understood mathematical discipline. Another important advance came in 1662 from a London draper named John Graunt, who showed that there were predictable patterns of longevity and death in a group, or cohort, of people of the same age, despite the uncertainty of the date of death of any one individual. This study became the basis for the original life table. One could now set up an insurance scheme to provide life insurance or pensions for a group of people, and to calculate with some degree of accuracy how much each person in the group should contribute to a common fund assumed to earn a fixed rate of interest. The first person to demonstrate publicly how this could be done was Edmond Halley (of Halley's comet fame). Halley constructed his own life table, and showed how it could be used to calculate the premium amount someone of a given age should pay to purchase a life annuity (Halley 1693).

Early Actuaries

James Dodson's pioneering work on the long term insurance contracts under which the same premium is charged each year led to the formation of the Society for Equitable Assurances on Lives and Survivorship (now commonly known as Equitable Life) in London in 1762 (Lewin 2007). Many other life insurance companies and pension funds were created over the following 200 years. Equitable Life was the first to use the word "actuary" for its chief executive officer in 1762 (Ogborn 1956). Previously, "actuary" meant an official who recorded the decisions, or "acts", of ecclesiastical courts (Faculty and Institute of Actuaries 2004). Other companies that did not use such mathematical and scientific methods most often failed or were forced to adopt the methods pioneered by Equitable (Bühlmann 1997).

Technological Advances

In the 18th and 19th centuries, calculations were of course performed without computers. The calculations of life insurance premiums and reserving requirements are rather complex, and actuaries developed techniques to make the calculations as easy as possible, for example "commutation functions" (essentially precalculated columns of summations over time of discounted values of survival and death probabilities) (Slud 2006). Actuarial organizations were founded to support and further both actuaries and actuarial science, and to protect the public interest by promoting competency and ethical standards (Hickman 2004). However, calculations remained cumbersome, and actuarial shortcuts were commonplace. Non-life actuaries followed in the footsteps of their life insurance colleagues during the 20th century. The 1920 revision for the New-York based National Council on Workmen's Compensation Insurance rates took over two months of around-the-clock work by day and night teams of actuaries (Michelbacher 1920). In the 1930s and 1940s, the mathematical foundations for stochastic processes were developed (Bühlmann 1997). Actuaries could now begin to estimate losses using models of random events, instead of the deterministic methods they had used in the past. The introduction and development of the computer further revolutionized the actuarial profession. From pencil-and-paper to punchcards to current high-speed devices, the modeling and forecasting ability of the actuary has rapidly improved, while still being heavily dependent on the assumptions input into the models, and actuaries needed to adjust to this new world (MacGinnitie 1980).

Actuarial Science Related to Modern Financial Economics

Traditional actuarial science and modern financial economics in the USA have different practices, which is caused by 1) different ways of calculating funding and investment strategies and 2) different regulations.

Regulations are from the Armstrong investigation of 1905, the Glass–Steagall Act of 1932, the adoption of the Mandatory Security Valuation Reserve by the National Association of Insurance Commissioners, which cushioned market fluctuations, and the Financial Accounting Standards Board, (FASB) in the USA and Canada, which regulates pensions valuations and funding.

History

Historically, much of the foundation of actuarial theory predated modern financial theory. In the early twentieth century, actuaries were developing many techniques that can be found in modern financial theory, but for various historical reasons, these developments did not achieve much recognition (Whelan 2002).

As a result, actuarial science developed along a different path, becoming more reliant on assumptions, as opposed to the arbitrage-free risk-neutral valuation concepts used in modern finance. The divergence is not related to the use of historical data and statistical projections of liability cash flows, but is instead caused by the manner in which traditional actuarial methods apply market data with those numbers. For example, one traditional actuarial method suggests that changing the asset allocation mix of investments can change the value of liabilities and assets (by changing the discount rate assumption). This concept is inconsistent with financial economics.

The potential of modern financial economics theory to complement existing actuarial science was recognized by actuaries in the mid-twentieth century (Bühlmann 1997,). In the late 1980s and early 1990s, there was a distinct effort for actuaries to combine financial theory and stochastic methods into their established models. (D'arcy 1989). Ideas from financial economics became increasingly influential in actuarial thinking, and actuarial science has started to embrace more sophisticated mathematical modelling of finance (Economist 2006). Today, the profession, both in practice and in the educational syllabi of many actuarial organizations, is cognizant of the need to reflect the combined approach of tables, loss models, stochastic methods, and financial theory (Feldblum 2001). However, assumption-dependent concepts are still widely used (such as the setting of the discount rate assumption as mentioned earlier), particularly in North America.

Product design adds another dimension to the debate. Financial economists argue that pension benefits are bond-like and should not be funded with equity investments without reflecting the risks of not achieving expected returns. But some pension products do reflect the risks of unexpected returns. In some cases, the pension beneficiary assumes the risk, or the employer assumes the risk. The current debate now seems to be focusing on four principles:

1. financial models should be free of arbitrage

2. assets and liabilities with identical cash flows should have the same price. This, of course, is at odds with FASB

3. the value of an asset is independent of its financing

4. the final issue deals with how pension assets should be invested.

Essentially, financial economics state that pension assets should not be invested in equities for a variety of theoretical and practical reasons. (Moriarty 2006).

Actuaries in Criminal Justice

There is an increasing trend to recognize that actuarial skills can be applied to a range of applications outside the traditional fields of insurance, pensions, etc. One notable example is the use in some US states of actuarial models to set criminal sentencing guidelines. These models attempt to predict the chance of re-offending according to rating factors which include the type of crime, age, educational background and ethnicity of the offender (Silver & Chow-Martin 2002). However, these models have been open to criticism as providing justification for discrimination against specific ethnic groups by law enforcement personnel. Whether this is statistically correct or a self-fulfilling correlation remains under debate (Harcourt 2003).

Another example is the use of actuarial models to assess the risk of sex offense recidivism. Actuarial models and associated tables, such as the MnSOST-R, Static-99, and SORAG, have been used since the late 1990s to determine the likelihood that a sex offender will re-offend and thus whether he or she should be institutionalized or set free (Nieto & Jung 2006).

Demography

Demography is the statistical study of populations, especially human beings. As a very general science, it can analyse any kind of dynamic living population, i.e., one that changes over time or space. Demography encompasses the study of the size, structure, and distribution of these populations, and spatial or temporal changes in them in response to birth, migration, ageing, and death. Based on the demographic research of the earth, earth's population up to the year 2050 and 2100 can be estimated by demographers. Demographics are quantifiable characteristics of a given population.

Demographic analysis can cover whole societies, or groups defined by criteria such as education, nationality, religion, and ethnicity. Educational institutions usually treat demography as a field of sociology, though there are a number of independent demography departments.

Formal demography limits its object of study to the measurement of population processes, while the broader field of social demography or population studies also analyses the relationships between economic, social, cultural, and biological processes influencing a population.

History

Demographic thoughts can be traced back to antiquity, and were present in many civilizations and cultures, like Ancient Greece, Ancient Rome, India and China. In ancient Greece, this can be found in the writings of Herodotus, Thucidides, Hippocrates, Epicurus, Protagoras, Polus, Plato and Aristotle. In Rome, writers and philosophers like Cicero, Seneca, Pliny the elder, Marcus Aurelius, Epictetus, Cato, and Collumella also expressed important ideas on this ground.

In the Middle ages, Christian thinkers devoted much time in refuting the Classical ideas on demography. Important contributors to the field were William of Conches, Bartholomew of Lucca, William of Auvergne, William of Pagula, and Ibn Khaldun.

One of the earliest demographic studies in the modern period was *Natural and Political Observations Made upon the Bills of Mortality* (1662) by John Graunt, which contains a primitive form of life table. Among the study's findings were that one third of the children in London died before their sixteenth birthday. Mathematicians, such as Edmond Halley, developed the life table as the basis for life insurance mathematics. Richard Price was credited with the first textbook on life contingencies published in 1771, followed later by Augustus de Morgan, 'On the Application of Probabilities to Life Contingencies' (1838).

At the end of the 18th century, Thomas Robert Malthus concluded that, if unchecked, populations would be subject to exponential growth. He feared that population growth would tend to outstrip growth in food production, leading to ever-increasing famine and poverty. He is seen as the intellectual father of ideas of overpopulation and the limits to growth. Later, more sophisticated and realistic models were presented by Benjamin Gompertz and Verhulst.

The period 1860-1910 can be characterized as a period of transition wherein demography emerged from statistics as a separate field of interest. This period included a panoply of international 'great demographers' like Adolphe Quételet (1796–1874), William Farr (1807–1883), Louis-Adolphe Bertillon (1821–1883) and his son Jacques (1851–1922), Joseph Körösi (1844–1906), Anders Nicolas Kaier (1838–1919), Richard Böckh (1824–1907), Émile Durkheim (1858-1917), Wilhelm Lexis (1837–1914) and Luigi Bodio (1840–1920) contributed to the development of demography and to the toolkit of methods and techniques of demographic analysis.

Methods

There are two types of data collection—direct and indirect—with several different methods of each type.

Direct Methods

Direct data comes from vital statistics registries that track all births and deaths as well as certain changes in legal status such as marriage, divorce, and migration (registration of place of residence). In developed countries with good registration systems (such as the United States and much of Europe), registry statistics are the best method for estimating the number of births and deaths.

A census is the other common direct method of collecting demographic data. A census is usually conducted by a national government and attempts to enumerate every person in a country. However, in contrast to vital statistics data, which are typically collected continuously and summarized on an annual basis, censuses typically occur only every 10 years or so, and thus are not usually the best source of data on births and deaths. Analyses are conducted after a census to estimate how much over or undercounting took place. These compare the sex ratios from the census data to those estimated from natural values and mortality data.

Censuses do more than just count people. They typically collect information about families or households in addition to individual characteristics such as age, sex, marital status, literacy/education, employment status, and occupation, and geographical location. They may also collect data on migration (or place of birth or of previous residence), language, religion, nationality (or ethnicity or race), and citizenship. In countries in which the vital registration system may be incomplete, the censuses are also used as a direct source of information about fertility and mortality; for example the censuses of the People's Republic of China gather information on births and deaths that occurred in the 18 months immediately preceding the census.

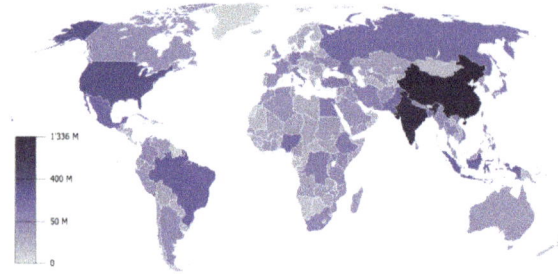

Map of countries by population

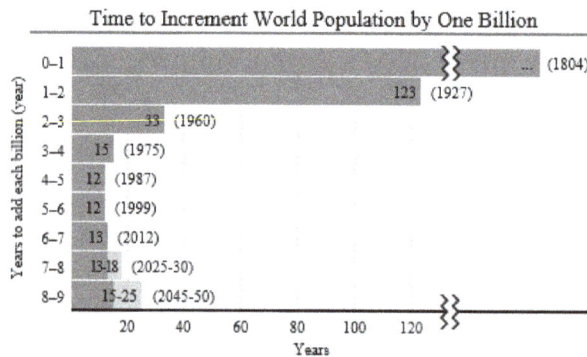

Rate of human population growth showing projections for later this century

Indirect Methods

Indirect methods of collecting data are required in countries and periods where full data are not available, such as is the case in much of the developing world, and most of historical demography. One of these techniques in contemporary demography is the sister method, where survey researchers ask women how many of their sisters have died or had children and at what age. With these surveys, researchers can then indirectly estimate birth or death rates for the entire population. Other indirect methods in contemporary demography include asking people about siblings, parents, and children. Other indirect methods are necessary in historical demography.

There are a variety of demographic methods for modelling population processes. They include models of mortality (including the life table, Gompertz models, hazards models, Cox proportional hazards models, multiple decrement life tables, Brass relational logits), fertility (Hernes model, Coale-Trussell models, parity progression ratios), marriage (Singulate Mean at Marriage, Page model), disability (Sullivan's method, multistate life tables), population projections (Lee Carter, the Leslie Matrix), and population momentum (Keyfitz).

The United Kingdom has a series of four national birth cohort studies, the first three spaced apart by 12 years: the 1946 National Survey of Health and Development, the 1958 National Child Development Study, the 1970 British Cohort Study, and the Millennium Cohort Study, begun much more recently in 2000. These have followed the lives of samples of people (typically beginning with around 17,000 in each study) for many years, and are still continuing. As the samples have been drawn in a nationally representative way, inferences can be drawn from these studies about the differences between four distinct generations of British people in terms of their health, education, attitudes, childbearing and employment patterns.

Common Rates and Ratios

- The crude birth rate, the annual number of live births per 1,000 people.

- The general fertility rate, the annual number of live births per 1,000 women of childbearing age (often taken to be from 15 to 49 years old, but sometimes from 15 to 44).

- The age-specific fertility rates, the annual number of live births per 1,000 women in particular age groups (usually age 15-19, 20-24 etc.)

- The crude death rate, the annual number of deaths per 1,000 people.

- The infant mortality rate, the annual number of deaths of children less than 1 year old per 1,000 live births.

- The expectation of life (or life expectancy), the number of years which an individual at a given age could expect to live at present mortality levels.

- The total fertility rate, the number of live births per woman completing her reproductive life, if her childbearing at each age reflected current age-specific fertility rates.

- The replacement level fertility, the average number of children women must have in order to replace the population for the next generation. For example, the replacement level fertility in the US is 2.11.

- The gross reproduction rate, the number of daughters who would be born to a woman completing her reproductive life at current age-specific fertility rates.

- The net reproduction ratio is the expected number of daughters, per newborn prospective mother, who may or may not survive to and through the ages of childbearing.

- A stable population, one that has had constant crude birth and death rates for such a long period of time that the percentage of people in every age class remains constant, or equivalently, the population pyramid has an unchanging structure.

- A stationary population, one that is both stable and unchanging in size (the difference between crude birth rate and crude death rate is zero).

A stable population does not necessarily remain fixed in size. It can be expanding or shrinking.

Note that the crude death rate as defined above and applied to a whole population can give a misleading impression. For example, the number of deaths per 1,000 people can be higher for developed nations than in less-developed countries, despite standards of health being better in developed countries. This is because developed countries have proportionally more older people, who are more likely to die in a given year, so that the overall mortality rate can be higher even if the mortality rate at any given age is lower. A more complete picture of mortality is given by a life table which summarizes mortality separately at each age. A life table is necessary to give a good estimate of life expectancy.

Basic Equation

Suppose that a country (or other entity) contains $Population_t$ persons at time t. What is the size of the population at time $t + 1$?

$$Population_{t+1} = Population_t + Naturalincrease_t + Netmigration_t.$$

Natural increase from time t to $t + 1$:

$$Naturalincrease_t = Births_t - Deaths_t$$

Net migration from time t to $t + 1$:

$$Netmigration_t = Immigration_t - Emigration_t$$

This basic equation can also be applied to subpopulations. For example, the population size of ethnic groups or nationalities within a given society or country is subject to the same sources of change. However, when dealing with ethnic groups, "net migration" might have to be subdivided into physical migration and ethnic reidentification (assimilation). Individuals who change their ethnic self-labels or whose ethnic classification in government statistics changes over time may be thought of as migrating or moving from one population subcategory to another.

More generally, while the basic demographic equation holds true by definition, in practice the recording and counting of events (births, deaths, immigration, emigration) and the enumeration of the total population size are subject to error. So allowance needs to be made for error in the underlying statistics when any accounting of population size or change is made.

The figure in this section shows the latest (2004) UN projections of world population out to the year 2150 (red = high, orange = medium, green = low). The UN "medium" projection shows world population reaching an approximate equilibrium at 9 billion by 2075. Working independently, demographers at the International Institute for Applied Systems Analysis in Austria expect world population to peak at 9 billion by 2070. Throughout the 21st century, the average age of the population is likely to continue to rise.

Science of Population

Populations can change through three processes: fertility, mortality, and migration. Fertility involves the number of children that women have and is to be contrasted with fecundity (a woman's childbearing potential). Mortality is the study of the causes, consequences, and measurement of processes affecting death to members of the population. Demographers most commonly study mortality using the Life Table, a statistical device which provides information about the mortality conditions (most notably the life expectancy) in the population.

Migration refers to the movement of persons from a locality of origin to a destination place across some pre-defined, political boundary. Migration researchers do not designate movements 'migrations' unless they are somewhat permanent. Thus demographers do not consider tourists and travellers to be migrating. While demographers who study migration typically do so through census data on place of residence, indirect sources of data including tax forms and labour force surveys are also important.

Demography is today widely taught in many universities across the world, attracting students with initial training in social sciences, statistics or health studies. Being at the crossroads of several disciplines such as sociology, economics, epidemiology, geography, anthropology and history, demography offers tools to approach a large range of population issues by combining a more technical quantitative approach that represents the core of the discipline with many other methods borrowed from social or other sciences. Demographic research is conducted in universities, in research institutes as well as in statistical departments and in several international agencies. Population institutions are part of the Cicred (International Committee for Coordination of Demographic Research) network while most individual scientists engaged in demographic research are members of the International Union for the Scientific Study of Population, or a national association such as the Population Association of America in the United States, or affiliates of the Federation of Canadian Demographers in Canada.

Environmental Statistics

Environment statistics is the application of statistical methods to environmental science. It covers procedures for dealing with questions concerning both the natural environment in its undisturbed state and the interaction of humanity with the environment. Thus weather, climate, air and water quality are included, as are studies of plant and animal populations.

The United Nations Framework for the Development of Environment Statistics

(FDES) defines the scope of environment statistics as follows: The scope of environment statistics covers biophysical aspects of the environment and those aspects of the socio-economic system that directly influence and interact with the environment. The scope of environment, social and economic statistics overlap. It is not easy – or necessary – to draw a clear line dividing these areas. Social and economic statistics that describe processes or activities with a direct impact on, or direct interaction with, the environment are used widely in environment statistics. They are within the scope of the FDES.

Environmental statistics covers a number of types of study:

- Baseline studies to document the present state of an environment to provide background in case of unknown changes in the future;

- Targeted studies to describe the likely impact of changes being planned or of accidental occurrences;

- Regular monitoring to attempt to detect changes in the environment.

Economic Statistics

Economic statistics is a topic in applied statistics that concerns the collection, processing, compilation, dissemination, and analysis of economic data. It is also common to call the data themselves 'economic statistics', but for this usage see economic data. The data of concern to economic statistics may include those of an economy of region, country, or group of countries. Economic statistics may also refer to a subtopic of official statistics for data produced by official organizations (e.g. national statistical services, intergovernmental organizations such as United Nations, European Union or OECD, central banks, ministries, etc.). Analyses within economic statistics both make use of and provide the empirical data needed in economic research, whether descriptive or econometric. They are a key input for decision making as to economic policy. The subject includes statistical analysis of topics and problems in microeconomics, macroeconomics, business, finance, forecasting, data quality, and policy evaluation. It also includes such considerations as what data to collect in order to quantify some particular aspect of an economy and of how best to collect in any given instance.

Statistical Process Control

Statistical process control (SPC) is a method of quality control which uses statistical methods. SPC is applied in order to monitor and control a process. Monitoring and controlling the process ensures that it operates at its full potential. At its full potential, the process can make as much conforming product as possible with a minimum (if not an elimination) of waste (rework or scrap). SPC can be applied to any process where the "conforming product" (product meeting specifications) output can be measured. Key tools used in SPC include control charts; a focus on continuous improvement; and the design of experiments. An example of a process where SPC is applied is manufacturing lines.

Overview

Objective Analysis

SPC must be practiced in 2 phases: The first phase is the initial establishment of the process, and the second phase is the regular production use of the process. In the second phase, a decision of the period to be examined must be made, depending upon the change in 4 - M conditions (Man, Machine, Material, Method) and wear rate of parts used in the manufacturing process (machine parts, jigs, and fixture)

Emphasis on Early Detection

An advantage of SPC over other methods of quality control, such as "inspection", is that it emphasizes early detection and prevention of problems, rather than the correction of problems after they have occurred.

Increasing rate of Production

In addition to reducing waste, SPC can lead to a reduction in the time required to produce the product. SPC makes it less likely the finished product will need to be reworked.

Limitations

SPC is applied to reduce or eliminate process waste. This, in turn, eliminates the need for the process step of post-manufacture inspection. The success of SPC relies not only on the skill with which it is applied, but also on how suitable or amenable the process is to SPC. In some cases, it may be difficult to judge when the application of SPC is appropriate.

History

SPC was pioneered by Walter A. Shewhart at Bell Laboratories in the early 1920s. Shewhart developed the control chart in 1924 and the concept of a state of statistical control. Statistical control is equivalent to the concept of exchangeability developed by logician William Ernest Johnson also in 1924 in his book *Logic, Part III: The Logical Foundations of Science*. Along with a gifted team at AT&T that included Harold Dodge and Harry Romig he worked to put sampling inspection on a rational statistical basis as well. Shewhart consulted with Colonel Leslie E. Simon in the application of control charts to munitions manufacture at the Army's Picatinny Arsenal in 1934. That successful application helped convince Army Ordnance to engage AT&T's George Edwards to consult on the use of statistical quality control among its divisions and contractors at the outbreak of World War II.

W. Edwards Deming invited Shewhart to speak at the Graduate School of the U.S. Department of Agriculture, and served as the editor of Shewhart's book *Statistical Method from the Viewpoint of Quality Control* (1939) which was the result of that lecture. Deming was an important architect of the quality control short courses that trained American industry in the new techniques during WWII. The graduates of these wartime courses formed a new professional society in 1945, the American Society for Quality Control, which elected Edwards as its first president. Deming traveled to Japan during the Allied Occupation and met with the Union of Japanese Scientists and Engineers (JUSE) in an effort to introduce SPC methods to Japanese industry .

"Common" and "Special" Sources of Variation

Shewhart read the new statistical theories coming out of Britain, especially the work of William Sealy Gosset, Karl Pearson, and Ronald Fisher. However, he understood that data from physical processes seldom produced a "normal distribution curve"; that is, a Gaussian distribution or "bell curve". He discovered that data from measurements of variation in manufacturing did not always behave the way as data from measurements of natural phenomena (for example, Brownian motion of particles). Shewhart concluded that while every process displays variation, some processes display variation that is natural to the process ("common" sources of variation)- these processes were described as 'in (statistical) control'. Other processes additionally display variation that is not present in the causal system of the process at all times ("special" sources of variation), and these were described as 'not in control'.

Application to Non-manufacturing Processes

In 1988, the Software Engineering Institute suggested that SPC could be applied to non-manufacturing processes, such as software engineering processes, in the Capability Maturity Model (CMM). The Level 4 and Level 5 practices of the Capability Maturity Model Integration (CMMI) use this concept.

The notion that SPC is a useful tool when applied to non-repetitive, knowledge-intensive processes such as research and development or systems engineering has encountered skepticism and remains controversial.

In his seminal article No Silver Bullet, Fred Brooks points out that the complexity, conformance requirements, changeability, and invisibility of software results in inherent and essential variation that cannot be removed. This implies that SPC is less effective in the domain of software development than in, e.g., manufacturing.

In 2014 a method for data validation of measurement data, based on SPC, was tried out. The method enabled the user to validate data containing static wave components (process noise), a requirement when working on hydro power plants where slowly damping surges are abundant during normal operation.

Variation in Manufacturing

In manufacturing, quality is defined as conformance to specification. However, no two products or characteristics are ever exactly the same, because any process contains many sources of variability. In mass-manufacturing, traditionally, the quality of a finished article is ensured by post-manufacturing inspection of the product. Each article may be accepted or rejected according to how well it meets its design specifications. In contrast, SPC uses statistical tools to observe the performance of the production process in order to detect significant variations before they result in the production of a sub-standard article. Any source of variation at any point of time in a process will fall into one of two classes.

1) "Common Causes" - sometimes referred to as nonassignable, normal sources of variation. It refers to many sources of variation that consistently acts on process. These types of causes produce a stable and repeatable distribution over time.

2) "Special Causes" - sometimes referred to as assignable sources of variation. It refers to any factor causing variation that affects only some of the process output. They are often intermittent and unpredictable.

Most processes have many sources of variation; most of them are minor and may be ignored. If the dominant sources of variation are identified, however, resources for change can be focused on them. If the dominant assignable sources of variation are detected, potentially they can be identified and removed. Once removed, the process is said to be "stable". When a process is stable, its variation should remain within a known set of limits. That is, at least, until another assignable source of variation occurs. For example, a breakfast cereal packaging line may be designed to fill each cereal box with 500 grams of cereal. Some boxes will have slightly more than 500 grams, and some will have slightly less. When the package weights are measured, the data will demonstrate a distribution of net weights. If the production process, its inputs, or its environment (for example, the machines on the line) change, the distribution of the data will change. For example, as the cams and pulleys of the machinery wear, the cereal filling machine may put more than the specified amount of cereal into each box. Although this might benefit the customer, from the manufacturer's point of view, this is wasteful and increases the cost of production. If the manufacturer finds the change and its source in a timely manner, the change can be corrected (for example, the cams and pulleys replaced).

Application of SPC

The application of SPC involves three main phases of activity:

1. Understanding the process and the specification limits.

2. Eliminating assignable (special) sources of variation, so that the process is stable.

3. Monitoring the ongoing production process, assisted by the use of control charts, to detect significant changes of mean or variation.

Control Charts

The data from measurements of variations at points on the process map is monitored using control charts. Control charts attempt to differentiate "assignable" ("special") sources of variation from "common" sources. "Common" sources, because they are an expected part of the process, are of much less concern to the manufacturer than "assignable" sources. Using control charts is a continuous activity, ongoing over time.

Stable Process

When the process does not trigger any of the control chart "detection rules" for the control chart, it is said to be "stable". A process capability analysis may be performed on a stable process to predict the ability of the process to produce "conforming product" in the future.

Excessive Variation

When the process triggers any of the control chart "detection rules", (or alternatively, the process capability is low), other activities may be performed to identify the source of the excessive variation. The tools used in these extra activities include: Ishikawa diagram, designed experiments, and Pareto charts. Designed experiments are a means of objectively quantifying the relative importance (strength) of sources of variation. Once the sources of (special cause) variation are identified, they can be minimized or eliminated. Steps to eliminating a source of variation might include: development of standards, staff training, error-proofing, and changes to the process itself or its inputs.

Mathematics of Control Charts

Digital control charts use logic-based rules that determine "derived values" which signal the need for correction. For example, derived value = last value + average absolute difference between the last N numbers.

Statistical Mechanics

Statistical mechanics is a branch of theoretical physics that studies, using probability theory, the average behaviour of a mechanical system where the state of the system is uncertain.

A common use of statistical mechanics is in explaining the thermodynamic behaviour of large systems. This branch of statistical mechanics which treats and extends classical thermodynamics is known as statistical thermodynamics or equilibrium statistical mechanics. Microscopic mechanical laws do not contain concepts such as temperature, heat, or entropy; however, statistical mechanics shows how these concepts arise from the natural uncertainty about the state of a system when that system is prepared in practice. The benefit of using statistical mechanics is that it provides exact methods to connect thermodynamic quantities (such as heat capacity) to microscopic behaviour, whereas, in classical thermodynamics, the only available option would be to just measure and tabulate such quantities for various materials. Statistical mechanics also makes it possible to *extend* the laws of thermodynamics to cases which are not considered in classical thermodynamics, such as microscopic systems and other mechanical systems with few degrees of freedom.

Statistical mechanics also finds use outside equilibrium. An important subbranch known as non-equilibrium statistical mechanics deals with the issue of microscopically modelling the speed of irreversible processes that are driven by imbalances. Examples of such processes include chemical reactions or flows of particles and heat. Unlike with equilibrium, there is no exact formalism that applies to non-equilibrium statistical mechanics in general, and so this branch of statistical mechanics remains an active area of theoretical research.

Principles: Mechanics And Ensembles

In physics there are two types of mechanics usually examined: classical mechanics and quantum mechanics. For both types of mechanics, the standard mathematical approach is to consider two concepts:

1. The complete state of the mechanical system at a given time, mathematically encoded as a phase point (classical mechanics) or a pure quantum state vector (quantum mechanics).

2. An equation of motion which carries the state forward in time: Hamilton's equations (classical mechanics) or the time-dependent Schrödinger equation (quantum mechanics)

Using these two concepts, the state at any other time, past or future, can in principle be calculated. There is however a disconnection between these laws and everyday life experiences, as we do not find it necessary (nor even theoretically possible) to know exactly at a microscopic level the simultaneous positions and velocities of each molecule while carrying out processes at the human scale (for example, when performing a chemical reaction). Statistical mechanics fills this disconnection between the laws of mechanics and the practical experience of incomplete knowledge, by adding some uncertainty about which state the system is in.

Whereas ordinary mechanics only considers the behaviour of a single state, statistical mechanics introduces the statistical ensemble, which is a large collection of virtual, independent copies of the system in various states. The statistical ensemble is a probability distribution over all possible states of the system. In classical statistical mechanics, the ensemble is a probability distribution over phase points (as opposed to a single phase point in ordinary mechanics), usually represented as a distribution in a phase space with canonical coordinates. In quantum statistical mechanics, the ensemble is a probability distribution over pure states, and can be compactly summarized as a density matrix.

As is usual for probabilities, the ensemble can be interpreted in different ways:

- an ensemble can be taken to represent the various possible states that a *single system* could be in (epistemic probability, a form of knowledge), or

- the members of the ensemble can be understood as the states of the systems in experiments repeated on independent systems which have been prepared in a similar but imperfectly controlled manner (empirical probability), in the limit of an infinite number of trials.

These two meanings are equivalent for many purposes, and will be used interchangeably in this chapter.

However the probability is interpreted, each state in the ensemble evolves over time according to the equation of motion. Thus, the ensemble itself (the probability distribution over states) also evolves, as the virtual systems in the ensemble continually leave one state and enter another. The ensemble evolution is given by the Liouville equation (classical mechanics) or the von Neumann equation (quantum mechanics). These equations are simply derived by the application of the mechanical equation of motion separately to each virtual system contained in the ensemble, with the probability of the virtual system being conserved over time as it evolves from state to state.

One special class of ensemble is those ensembles that do not evolve over time. These ensembles are known as *equilibrium ensembles* and their condition is known as *statistical equilibrium*. Statistical equilibrium occurs if, for each state in the ensemble, the ensemble also contains all of its future and past states with probabilities equal to the probability of being in that state. The study of equilibrium ensembles of isolated systems is the focus of statistical thermodynamics. Non-equilibrium statistical mechanics addresses the more general case of ensembles that change over time, and/or ensembles of non-isolated systems.

Statistical Thermodynamics

The primary goal of statistical thermodynamics (also known as equilibrium statistical mechanics) is to derive the classical thermodynamics of materials in terms of the properties of their constituent particles and the interactions between them. In other words, statistical thermodynamics provides a connection between the macroscopic properties of materials in thermodynamic equilibrium, and the microscopic behaviours and motions occurring inside the material.

Whereas statistical mechanics proper involves dynamics, here the attention is focussed on *statistical equilibrium* (steady state). Statistical equilibrium does not mean that the particles have stopped moving (mechanical equilibrium), rather, only that the ensemble is not evolving.

Fundamental Postulate

A sufficient (but not necessary) condition for statistical equilibrium with an isolated system is that the probability distribution is a function only of conserved properties (total energy, total particle numbers, etc.). There are many different equilibrium ensembles that can be considered, and only some of them correspond to thermodynamics. Additional postulates are necessary to motivate why the ensemble for a given system should have one form or another.

A common approach found in many textbooks is to take the *equal a priori probability postulate*. This postulate states that

> *For an isolated system with an exactly known energy and exactly known composition, the system can be found with* equal probability *in any microstate consistent with that knowledge.*

The equal a priori probability postulate therefore provides a motivation for the microcanonical ensemble described below. There are various arguments in favour of the equal a priori probability postulate:

- Ergodic hypothesis: An ergodic state is one that evolves over time to explore "all accessible" states: all those with the same energy and composition. In an ergodic system, the microcanonical ensemble is the only possible equilibrium ensemble with fixed energy. This approach has limited applicability, since most systems are not ergodic.

- Principle of indifference: In the absence of any further information, we can only assign equal probabilities to each compatible situation.

- Maximum information entropy: A more elaborate version of the principle of indifference states that the correct ensemble is the ensemble that is compatible with the known information and that has the largest Gibbs entropy (information entropy).

Other fundamental postulates for statistical mechanics have also been proposed.

Three Thermodynamic Ensembles

There are three equilibrium ensembles with a simple form that can be defined for any isolated system bounded inside a finite volume. These are the most often discussed ensembles in statistical thermodynamics. In the macroscopic limit (defined below) they all correspond to classical thermodynamics.

Microcanonical ensemble

> describes a system with a precisely given energy and fixed composition (precise number of particles). The microcanonical ensemble contains with equal probability each possible state that is consistent with that energy and composition.

Canonical ensemble

> describes a system of fixed composition that is in thermal equilibrium with a heat bath of a precise temperature. The canonical ensemble contains states of varying energy but identical composition; the different states in the ensemble are accorded different probabilities depending on their total energy.

Grand canonical ensemble

> describes a system with non-fixed composition (uncertain particle numbers) that is in thermal and chemical equilibrium with a thermodynamic reservoir. The reservoir has a precise temperature, and precise chemical potentials for various types of particle. The grand canonical ensemble contains states of varying energy and varying numbers of particles; the different states in the ensemble are accorded different probabilities depending on their total energy and total particle numbers.

For systems containing many particles (the thermodynamic limit), all three of the ensembles listed above tend to give identical behaviour. It is then simply a matter of mathematical convenience which ensemble is used.

Important cases where the thermodynamic ensembles *do not* give identical results include:

- Microscopic systems.

- Large systems at a phase transition.

- Large systems with long-range interactions.

In these cases the correct thermodynamic ensemble must be chosen as there are observable differences between these ensembles not just in the size of fluctuations, but also in average quantities such as the distribution of particles. The correct ensemble is that which corresponds to the way the system has been prepared and characterized—in other words, the ensemble that reflects the knowledge about that system.

	Thermodynamic ensembles		
	Microcanonical	**Canonical**	**Grand canonical**
Fixed variables	*N, E, V*	*N, T, V*	*μ, T, V*
Microscopic features	• Number of microstates	• Canonical partition function	• Grand partition function
Macroscopic function	• Boltzmann entropy	• Helmholtz free energy	• Grand potential

Calculation Methods

Once the characteristic state function for an ensemble has been calculated for a given system, that system is 'solved' (macroscopic observables can be extracted from the characteristic state function). Calculating the characteristic state function of a thermodynamic ensemble is not necessarily a simple task, however, since it involves considering every possible state of the system. While some hypothetical systems have been exactly solved, the most general (and realistic) case is too complex for exact solution. Various approaches exist to approximate the true ensemble and allow calculation of average quantities.

Exact

There are some cases which allow exact solutions.

- For very small microscopic systems, the ensembles can be directly computed by simply enumerating over all possible states of the system (using exact diagonalization in quantum mechanics, or integral over all phase space in classical mechanics).

- Some large systems consist of many separable microscopic systems, and each of the subsystems can be analysed independently. Notably, idealized gases of non-interacting particles have this property, allowing exact derivations of Max-well–Boltzmann statistics, Fermi–Dirac statistics, and Bose–Einstein statistics.

- A few large systems with interaction have been solved. By the use of subtle mathematical techniques, exact solutions have been found for a few toy mod-els. Some examples include the Bethe ansatz, square-lattice Ising model in zero field, hard hexagon model.

Monte Carlo

One approximate approach that is particularly well suited to computers is the Monte Carlo method, which examines just a few of the possible states of the system, with the states chosen randomly (with a fair weight). As long as these states form a represen-tative sample of the whole set of states of the system, the approximate characteristic function is obtained. As more and more random samples are included, the errors are reduced to an arbitrarily low level.

- The Metropolis–Hastings algorithm is a classic Monte Carlo method which was initially used to sample the canonical ensemble.

- Path integral Monte Carlo, also used to sample the canonical ensemble.

Other

- For rarefied non-ideal gases, approaches such as the cluster expansion use per-turbation theory to include the effect of weak interactions, leading to a virial expansion.

- For dense fluids, another approximate approach is based on reduced distribu-tion functions, in particular the radial distribution function.

- Molecular dynamics computer simulations can be used to calculate microca-nonical ensemble averages, in ergodic systems. With the inclusion of a connec-tion to a stochastic heat bath, they can also model canonical and grand canon-ical conditions.

- Mixed methods involving non-equilibrium statistical mechanical results may be useful.

Non-equilibrium Statistical Mechanics

There are many physical phenomena of interest that involve quasi-thermodynamic processes out of equilibrium, for example:

- heat transport by the internal motions in a material, driven by a temperature imbalance,

- electric currents carried by the motion of charges in a conductor, driven by a voltage imbalance,

- spontaneous chemical reactions driven by a decrease in free energy,

- friction, dissipation, quantum decoherence,

- systems being pumped by external forces (optical pumping, etc.),

- and irreversible processes in general.

All of these processes occur over time with characteristic rates, and these rates are of importance for engineering. The field of non-equilibrium statistical mechanics is concerned with understanding these non-equilibrium processes at the microscopic level. (Statistical thermodynamics can only be used to calculate the final result, after the external imbalances have been removed and the ensemble has settled back down to equilibrium.)

In principle, non-equilibrium statistical mechanics could be mathematically exact: ensembles for an isolated system evolve over time according to deterministic equations such as Liouville's equation or its quantum equivalent, the von Neumann equation. These equations are the result of applying the mechanical equations of motion independently to each state in the ensemble. Unfortunately, these ensemble evolution equations inherit much of the complexity of the underlying mechanical motion, and so exact solutions are very difficult to obtain. Moreover, the ensemble evolution equations are fully reversible and do not destroy information (the ensemble's Gibbs entropy is preserved). In order to make headway in modelling irreversible processes, it is necessary to consider additional factors besides probability and reversible mechanics.

Non-equilibrium mechanics is therefore an active area of theoretical research as the range of validity of these additional assumptions continues to be explored. A few approaches are described in the following subsections.

Stochastic Methods

One approach to non-equilibrium statistical mechanics is to incorporate stochastic (random) behaviour into the system. Stochastic behaviour destroys information contained in the ensemble. While this is technically inaccurate (aside from hypothetical situations involving black holes, a system cannot in itself cause loss of information), the randomness is added to reflect that information of interest becomes converted over time into subtle correlations within the system, or to correlations between the system and environment. These correlations appear as chaotic or pseudorandom influences on the variables of interest. By replacing these correlations with randomness proper, the calculations can be made much easier.

- *Boltzmann transport equation*: An early form of stochastic mechanics appeared even before the term "statistical mechanics" had been coined, in studies of kinetic theory. James Clerk Maxwell had demonstrated that molecular collisions would lead to apparently chaotic motion inside a gas. Ludwig Boltzmann subsequently showed that, by taking this molecular chaos for granted as a complete randomization, the motions of particles in a gas would follow a simple Boltzmann transport equation that would rapidly restore a gas to an equilibrium state.

  The Boltzmann transport equation and related approaches are important tools in non-equilibrium statistical mechanics due to their extreme simplicity. These approximations work well in systems where the "interesting" information is immediately (after just one collision) scrambled up into subtle correlations, which essentially restricts them to rarefied gases. The Boltzmann transport equation has been found to be very useful in simulations of electron transport in lightly doped semiconductors (in transistors), where the electrons are indeed analogous to a rarefied gas.

  A quantum technique related in theme is the random phase approximation.

- *BBGKY hierarchy*: In liquids and dense gases, it is not valid to immediately discard the correlations between particles after one collision. The BBGKY hierarchy (Bogoliubov–Born–Green–Kirkwood–Yvon hierarchy) gives a method for deriving Boltzmann-type equations but also extending them beyond the dilute gas case, to include correlations after a few collisions.

- *Keldysh formalism* (a.k.a. NEGF—non-equilibrium Green functions): A quantum approach to including stochastic dynamics is found in the Keldysh formalism. This approach often used in electronic quantum transport calculations.

Near-equilibrium Methods

Another important class of non-equilibrium statistical mechanical models deals with systems that are only very slightly perturbed from equilibrium. With very small perturbations, the response can be analysed in linear response theory. A remarkable result, as formalized by the fluctuation-dissipation theorem, is that the response of a system when near equilibrium is precisely related to the fluctuations that occur when the system is in total equilibrium. Essentially, a system that is slightly away from equilibrium—whether put there by external forces or by fluctuations—relaxes towards equilibrium in the same way, since the system cannot tell the difference or "know" how it came to be away from equilibrium.

This provides an indirect avenue for obtaining numbers such as ohmic conductivity and thermal conductivity by extracting results from equilibrium statistical mechanics. Since equilibrium statistical mechanics is mathematically well defined and (in some cases) more amenable for calculations, the fluctuation-dissipation connection can be a convenient shortcut for calculations in near-equilibrium statistical mechanics.

A few of the theoretical tools used to make this connection include:

- Fluctuation–dissipation theorem
- Onsager reciprocal relations
- Green–Kubo relations
- Landauer–Büttiker formalism
- Mori–Zwanzig formalism

Hybrid Methods

An advanced approach uses a combination of stochastic methods and linear response theory. As an example, one approach to compute quantum coherence effects (weak localization, conductance fluctuations) in the conductance of an electronic system is the use of the Green-Kubo relations, with the inclusion of stochastic dephasing by interactions between various electrons by use of the Keldysh method.

Applications Outside Thermodynamics

The ensemble formalism also can be used to analyze general mechanical systems with uncertainty in knowledge about the state of a system. Ensembles are also used in:

- propagation of uncertainty over time,
- regression analysis of gravitational orbits,
- ensemble forecasting of weather,
- dynamics of neural networks,
- bounded-rational potential games in game theory and economics.

History

In 1738, Swiss physicist and mathematician Daniel Bernoulli published *Hydrodynamica* which laid the basis for the kinetic theory of gases. In this work, Bernoulli posited the argument, still used to this day, that gases consist of great numbers of molecules moving in all directions, that their impact on a surface causes the gas pressure that we feel, and that what we experience as heat is simply the kinetic energy of their motion.

In 1859, after reading a paper on the diffusion of molecules by Rudolf Clausius, Scottish physicist James Clerk Maxwell formulated the Maxwell distribution of molecular velocities, which gave the proportion of molecules having a certain velocity in a specific range. This was the first-ever statistical law in physics. Five years later, in 1864, Ludwig Boltzmann, a young student in Vienna, came across Maxwell's paper and spent much of his life developing the subject further.

Statistical mechanics proper was initiated in the 1870s with the work of Boltzmann, much of which was collectively published in his 1896 *Lectures on Gas Theory*. Boltzmann's original papers on the statistical interpretation of thermodynamics, the H-theorem, transport theory, thermal equilibrium, the equation of state of gases, and similar subjects, occupy about 2,000 pages in the proceedings of the Vienna Academy and other societies. Boltzmann introduced the concept of an equilibrium statistical ensemble and also investigated for the first time non-equilibrium statistical mechanics, with his *H*-theorem.

The term "statistical mechanics" was coined by the American mathematical physicist J. Willard Gibbs in 1884. "Probabilistic mechanics" might today seem a more appropriate term, but "statistical mechanics" is firmly entrenched. Shortly before his death, Gibbs published in 1902 *Elementary Principles in Statistical Mechanics*, a book which formalized statistical mechanics as a fully general approach to address all mechanical systems—macroscopic or microscopic, gaseous or non-gaseous. Gibbs' methods were initially derived in the framework classical mechanics, however they were of such generality that they were found to adapt easily to the later quantum mechanics, and still form the foundation of statistical mechanics to this day.

References

- Balescu, Radu (1975). Equilibrium and Non-Equilibrium Statistical Mechanics. John Wiley & Sons. ISBN 9780471046004.

- Manly B.F.J. (2001) Statistics for Environmental Science and Management, Chapman & Hall/CRC. ISBN 1-58488-029-5

- Andrey Korotayev, Artemy Malkov, & Daria Khaltourina (2006). Introduction to Social Macrodynamics: Compact Macromodels of the World System Growth. Moscow: URSS, ISBN 5-484-00414-4.

- Perkins, Judith (August 25, 1995). The Suffering Self; Pain and Narrative Representation in the Early Christian Era. London: Routledge. ISBN 0-415-11363-6. LCCN 94042650.

- Fundamentals of Statistical and Thermal Physics. McGraw–Hill. p. 227. ISBN 9780070518001.

- Baxter, Rodney J. (1982). Exactly solved models in statistical mechanics. Academic Press Inc. ISBN 9780120831807.

- Mahon, Basil (2003). The Man Who Changed Everything – the Life of James Clerk Maxwell. Hoboken, NJ: Wiley. ISBN 0-470-86171-1. OCLC 52358254.

- "History of the actuarial profession". Faculty and Institute of Actuaries. 2004-01-13. Archived from the original on 2008-04-04. Retrieved 2010-09-26.

- Hsiao, William C (2004). "Harvard School of Public Health". Archived from the original (PDF) on 2007-03-27. Retrieved 2010-09-27.

Evolution of Statistics

Statistical methods date back to the 5th century BC. The early applications of statistical thinking dealt with the need of the state to base their policy on the economic data. The modern field of statistics emerged in the late 19th and 20th century. This chapter educates the reader with the history of statistics and progress it has made over a period of time.

History of Statistics

The History of statistics can be said to start around 1749 although, over time, there have been changes to the interpretation of the word *statistics*. In early times, the meaning was restricted to information about states. This was later extended to include all collections of information of all types, and later still it was extended to include the analysis and interpretation of such data. In modern terms, "statistics" means both sets of collected information, as in national accounts and temperature records, and analytical work which requires statistical inference.

Statistical activities are often associated with models expressed using probabilities, and require probability theory for them to be put on a firm theoretical basis.

A number of statistical concepts have had an important impact on a wide range of sciences. These include the design of experiments and approaches to statistical inference such as Bayesian inference, each of which can be considered to have their own sequence in the development of the ideas underlying modern statistics.

Introduction

By the 18th century, the term "statistics" designated the systematic collection of demographic and economic data by states. For at least two millennia, these data were mainly tabulations of human and material resources that might be taxed or put to military use. In the early 19th century, collection intensified, and the meaning of "statistics" broadened to include the discipline concerned with the collection, summary, and analysis of data. Today, data are collected and statistics are computed and widely distributed in government, business, most of the sciences and sports, and even for many pastimes. Electronic computers have expedited more elaborate statistical computation even as they have facilitated the collection and aggregation of data. A single data analyst may have available a set of data-files with millions of records, each with dozens or hundreds of separate measurements. These were collected over time from computer activity (for example, a stock exchange) or from computerized sensors, point-of-sale registers, and so on. Computers then produce simple, accurate summaries, and allow more tedious analyses, such as those that require inverting a large matrix or perform hundreds of steps of iteration, that would never be attempted by hand. Faster computing has allowed statisticians to develop "computer-intensive" methods which may look at all permutations, or use randomization to look at 10,000 permutations of a problem, to estimate answers that are not easy to quantify by theory alone.

The term "mathematical statistics" designates the mathematical theories of probability and statistical inference, which are used in statistical practice. The relation between statistics and probability theory developed rather late, however. In the 19th century, statistics increasingly used probability theory, whose initial results were found in the 17th and 18th centuries, particularly in the analysis of games of chance (gambling). By 1800, astronomy used probability models and statistical theories, particularly the method of least squares. Early probability theory and statistics was systematized in the 19th century and statistical reasoning and probability models were used by social scientists to advance the new sciences of experimental psychology and sociology, and by physical scientists in thermodynamics and statistical mechanics. The development of statistical reasoning was closely associated with the development of inductive logic and the scientific method, which are concerns that move statisticians away from the narrower area of mathematical statistics. Much of the theoretical work was readily available by the time computers were available to exploit them. By the 1970s, Johnson and Kotz produced a four-volume Compendium on Statistical Distributions (First Edition 1969-1972), which is still an invaluable resource.

Applied statistics can be regarded as not a field of mathematics but an autonomous mathematical science, like computer science and operations research. Unlike mathematics, statistics had its origins in public administration. Applications arose early in demography and economics; large areas of micro- and macro-economics today are "statistics" with an emphasis on time-series analyses. With its emphasis on learning from data and making best predictions, statistics also has been shaped by areas of academic research including psychological testing, medicine and epidemiology. The ideas of statistical testing have considerable overlap with decision science. With its concerns with searching and effectively presenting data, statistics has overlap with information science and computer science.

Etymology

The term *statistics* is ultimately derived from the New Latin *statisticum collegium* ("council of state") and the Italian word *statista* ("statesman" or "politician"). The German *Statistik*, first introduced by Gottfried Achenwall (1749), originally designated the analysis of data about the state, signifying the "science of state" (then called *political arithmetic* in English). It acquired the meaning of the collection and classification of data generally in the early 19th century. It was introduced into English in 1791 by Sir John Sinclair when he published the first of 21 volumes titled *Statistical Account of Scotland*.

Thus, the original principal purpose of *Statistik* was data to be used by governmental and (often centralized) administrative bodies. The collection of data about states and localities continues, largely through national and international statistical services. In particular, censuses provide frequently updated information about the population.

The first book to have 'statistics' in its title was "Contributions to Vital Statistics" (1845) by Francis GP Neison, actuary to the Medical Invalid and General Life Office.

Origins in Probability Theory

Basic forms of statistics have been used since the beginning of civilization. Early empires often collated censuses of the population or recorded the trade in various commodities. The Roman Empire was one of the first states to extensively gather data on the size of the empire's population, geographical area and wealth.

The use of statistical methods dates back to least to the 5th century BCE. The historian Thucydides in his *History of the Peloponnesian War* describes how the Athenians calculated the height of the wall of Platea by counting the number of bricks in an unplastered section of the wall sufficiently near them to be able to count them. The count was repeated several times by a number of soldiers. The most frequent value (in modern terminology - the mode) so determined was taken to be the most likely value of the number of bricks. Multiplying this value by the height of the bricks used in the wall allowed the Athenians to determine the height of the ladders necessary to scale the walls.

In the Indian epic - the Mahabharata (Book 3: The Story of Nala) - King Rtuparna estimated the number of fruit and leaves (2095 fruit and 50,000,000 - five crores - leaves) on two great branches of a Vibhitaka tree by counting them on a single twig. This number was then multiplied by the number of twigs on the branches. This estimate was later checked and found to be very close to the actual number. With knowledge of this method Nala was subsequently able to regain his kingdom.

The earliest writing on statistics was found in a 9th-century book entitled: "Manuscript on Deciphering Cryptographic Messages", written by Al-Kindi (801–873 CE). In his book, Al-Kindi gave a detailed description of how to use statistics and frequency analysis to decipher encrypted messages. This text arguably gave rise to the birth of both statistics and cryptanalysis.

The Trial of the Pyx is a test of the purity of the coinage of the Royal Mint which has been held on a regular basis since the 12th century. The Trial itself is based on statistical sampling methods. After minting a series of coins - originally from ten pounds of silver - a single coin was placed in the Pyx - a box in Westminster Abbey. After a given period - now once a year - the coins are removed and weighed. A sample of coins removed from the box are then tested for purity.

The *Nuova Cronica*, a 14th-century history of Florence by the Florentine banker and official Giovanni Villani, includes much statistical information on population, ordinances, commerce and trade, education, and religious facilities and has been described as the first introduction of statistics as a positive element in history, though neither the term nor the concept of statistics as a specific field yet existed. But this was proven to be incorrect after the rediscovery of Al-Kindi's book on frequency analysis.

The arithmetic mean, although a concept known to the Greeks, was not generalised to more than two values until the 16th century. The invention of the decimal system by Simon Stevin in 1585 seems likely to have facilitated these calculations. This method was first adopted in astronomy by Tycho Brahe who was attempting to reduce the errors in his estimates of the locations of various celestial bodies.

The idea of the median originated in Edward Wright's book on navigation (*Certaine Errors in Navigation*) in 1599 in a section concerning the determination of location with a compass. Wright felt that this value was the most likely to be the correct value in a series of observations.

Sir William Petty, a 17th-century economist who used early statistical methods to analyse demographic data.

The birth of statistics is often dated to 1662, when John Graunt, along with William Petty, developed early human statistical and census methods that provided a framework for modern demography. He produced the first life table, giving probabilities of survival to each age. His book *Natural and Political Observations Made upon the Bills of Mortality* used analysis of the mortality rolls to make the first statistically based estimation of the population of London. He knew that there were around 13,000 funerals per year in London and that three people died per eleven families per year. He estimated from the parish records that the average family size was 8 and calculated that the population of London was about 384,000. Laplace in 1802 estimated the population of France with a similar method.

Although the original scope of statistics was limited to data useful for governance, the approach was extended to many fields of a scientific or commercial nature during the 19th century. The mathematical foundations for the subject heavily drew on the new probability theory, pioneered in the 16th century in the correspondence amongst Gerolamo Cardano, Pierre de Fermat and Blaise Pascal. Christiaan Huygens (1657) gave the earliest known scientific treatment of the subject. Jakob Bernoulli's *Ars Conjectandi* (posthumous, 1713) and Abraham de Moivre's *The Doctrine of Chances* (1718) treated the subject as a branch of mathematics. In his book Bernoulli introduced the idea of representing complete certainty as one and probability as a number between zero and one.

The formal study of theory of errors may be traced back to Roger Cotes' *Opera Miscellanea* (posthumous, 1722), but a memoir prepared by Thomas Simpson in 1755 (printed 1756) first applied the theory to the discussion of errors of observation. The reprint (1757) of this memoir lays down the axioms that positive and negative errors are equally probable, and that there are certain assignable limits within which all errors may be supposed to fall; continuous errors are discussed and a probability curve is given. Simpson discussed several possible distributions of error. He first considered the uniform distribution and then the discrete symmetric triangular distribution followed by the continuous symmetric triangle distribution. Tobias Mayer, in his study of the libration of the moon (*Kosmographische Nachrichten*, Nuremberg, 1750), invented the first formal method for estimating the unknown quantities by generalized the averaging of observations under identical circumstances to the averaging of groups of similar equations.

Ruder Boškovic in 1755 based in his work on the shape of the earth proposed in his book *De Litteraria expeditione per pontificiam ditionem ad dimetiendos duos meridiani gradus a PP. Maire et Boscovicli* that the true value of a series of observations would be that which minimises the sum of absolute errors. In modern terminology this value is the median. The first example of what later became known as the normal curve was studied by Abraham de Moivre who plotted this curve on November 12, 1733. de Moivre was studying the number of heads that occurred when a 'fair' coin was tossed.

In 1761 Thomas Bayes proved Bayes' theorem and in 1765 Joseph Priestley invented the first timeline charts.

Johann Heinrich Lambert in his 1765 book *Anlage zur Architectonic* proposed the semicircle as a distribution of errors:

$$f(x) = \frac{1}{2}\sqrt{(1-x^2)}$$

with -1 < x < 1.

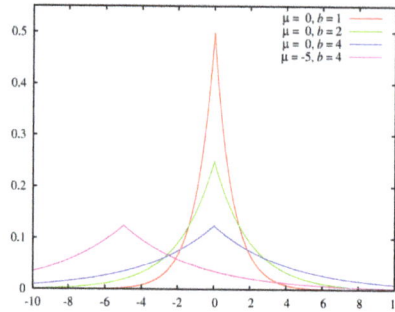

Probability density plots for the Laplace distribution.

Pierre-Simon Laplace (1774) made the first attempt to deduce a rule for the combination of observations from the principles of the theory of probabilities. He represented the law of probability of errors by a curve and deduced a formula for the mean of three observations.

Laplace in 1774 noted that the frequency of an error could be expressed as an exponential function of its magnitude once its sign was disregarded. This distribution is now known as the Laplace distribution. Lagrange proposed a parabolic distribution of errors in 1776.

Laplace in 1778 published his second law of errors wherein he noted that the frequency of an error was proportional to the exponential of the square of its magnitude. This was subsequently rediscovered by Gauss (possibly in 1795) and is now best known as the normal distribution which is of central importance in statistics. This distribution was first referred to as the *normal* distribution by Pierce in 1873 who was studying measurement errors when an object was dropped onto a wooden base. He chose the term *normal* because of its frequent occurrence in naturally occurring variables.

Lagrange also suggested in 1781 two other distributions for errors - a Raised cosine distribution and a logarithmic distribution.

Laplace gave (1781) a formula for the law of facility of error (a term due to Joseph Louis Lagrange, 1774), but one which led to unmanageable equations. Daniel Bernoulli (1778) introduced the principle of the maximum product of the probabilities of a system of concurrent errors.

In 1786 William Playfair (1759-1823) introduced the idea of graphical representation into statistics. He invented the line chart, bar chart and histogram and incorporated them into his works on economics, the *Commercial and Political Atlas*. This was followed in 1795 by his invention of the pie chart and circle chart which he used to display the evolution of England's imports and exports. These latter charts came to general attention when he published examples in his *Statistical Breviary* in 1801.

Laplace, in an investigation of the motions of Saturn and Jupiter in 1787, generalized Mayer's method by using different linear combinations of a single group of equations.

In 1802 Laplace estimated the population of France to be 28,328,612. He calculated this figure using the number of births in the previous year and census data for three communities. The census data of these communities showed that they had 2,037,615 persons and that the number of births were 71,866. Assuming that these samples were representative of France, Laplace produced his estimate for the entire population.

The method of least squares, which was used to minimize errors in data measurement, was published independently by Adrien-Marie Legendre (1805), Robert Adrain (1808), and Carl Friedrich Gauss (1809). Gauss had used the method in his famous 1801 prediction of the location of the dwarf planet Ceres. The observations that Gauss based his calculations on were made by the Italian monk Piazzi.

The term *probable error* (*der wahrscheinliche Fehler*) - the median deviation from the mean - was introduced in 1815 by the German astronomer Frederik Wilhelm Bessel. Antoine Augustin Cournot in 1843 was the first to use the term *median* (*valeur médiane*) for the value that divides a probability distribution into two equal halves.

Other contributors to the theory of errors were Ellis (1844), De Morgan (1864), Glaisher (1872), and Giovanni Schiaparelli (1875). Peters's (1856) formula for , the "probable error" of a single observation was widely used and inspired early robust statistics.

In the 19th century authors on statistical theory included Laplace, S. Lacroix (1816), Littrow (1833), Dedekind (1860), Helmert (1872), Laurent (1873), Liagre, Didion, De Morgan and Boole.

Gustav Theodor Fechner used the median (*Centralwerth*) in sociological and psychological phenomena. It had earlier been used only in astronomy and related fields. Francis Galton used the English term *median* for the first time in 1881 having earlier used the terms *middle-most value* in 1869 and the *medium* in 1880.

Adolphe Quetelet (1796–1874), another important founder of statistics, introduced the notion of the "average man" (*l'homme moyen*) as a means of understanding complex social phenomena such as crime rates, marriage rates, and suicide rates.

The first tests of the normal distribution were invented by the German statistician Wilhelm Lexis in the 1870s. The only data sets available to him that he was able to show were normally distributed were birth rates.

Development of Modern Statistics

Although the origins of statistical theory lie in the 18th century advances in probability, the modern field of statistics only emerged in the late 19th and early 20th century in three stages. The first wave, at the turn of the century, was led by the work of Francis Galton and Karl Pearson, who transformed statistics into a rigorous mathematical discipline used for analysis, not just in science, but in industry and politics as well. The second wave of the 1910s and 20s was initiated by William Gosset, and reached its culmination in the insights of Ronald Fisher. This involved the development of better design of experiments models, hypothesis testing and techniques for use with small data samples. The final wave, which mainly saw the refinement and expansion of earlier developments, emerged from the collaborative work between Egon Pearson and Jerzy Neyman in the 1930s. Today, statistical methods are applied in all fields that involve decision making, for making accurate inferences from a collated body of data and for making decisions in the face of uncertainty based on statistical methodology.

The original logo of the Royal Statistical Society, founded in 1834.

The first statistical bodies were established in the early 19th century. The Royal Statistical Society was founded in 1834 and Florence Nightingale, its first female member, pioneered the application of statistical analysis to health problems for the furtherance of epidemiological understanding and public health practice. However, the methods then used would not be considered as modern statistics today.

The Oxford scholar Francis Ysidro Edgeworth's book, *Metretike: or The Method of Measuring Probability and Utility* (1887) dealt with probability as the basis of inductive reasoning, and his later works focused on the 'philosophy of chance'. His first paper on statistics (1883) explored the law of error (normal distribution), and his *Methods of Statistics* (1885) introduced an early version of the t distribution, the Edgeworth expansion, the Edgeworth series, the method of variate transformation and the asymptotic theory of maximum likelihood estimates.

The Norwegian Anders Nicolai Kiær introduced the concept of stratified sampling in 1895. Arthur Lyon Bowley introduced new methods of data sampling in 1906 when working on social statistics. Although statistical surveys of social conditions had started with Charles Booth's "Life and Labour of the People in London" (1889-1903) and Seebohm Rowntree's "Poverty, A Study of Town Life" (1901), Bowley's, key innovation consisted of the use of random sampling techniques. His efforts culminated in his *New Survey of London Life and Labour*.

Francis Galton is credited as one of the principal founders of statistical theory. His contributions to the field included introducing the concepts of standard deviation, correlation, regression and the application of these methods to the study of the variety of human characteristics - height, weight, eyelash length among others. He found that many of these could be fitted to a normal curve distribution.

Galton submitted a paper to *Nature* in 1907 on the usefulness of the median. He examined the accuracy of 787 guesses of the weight of an ox at a country fair. The actual weight was 1208 pounds: the median guess was 1198. The guesses were markedly non-normally distributed.

Galton's publication of *Natural Inheritance* in 1889 sparked the interest of a brilliant mathematician, Karl Pearson, then working at University College London, and he went on to found the discipline of mathematical statistics. He emphasised the statistical foundation of scientific laws and promoted its study and his laboratory attracted students from around the world attracted by his new methods of analysis, including Udny Yule. His work grew to encompass the fields of biology, epidemiology, anthropometry, medicine and social history. In 1901, with Walter Weldon, founder of biometry, and Galton, he founded the journal *Biometrika* as the first journal of mathematical statistics and biometry.

His work, and that of Galton's, underpins many of the 'classical' statistical methods which are in common use today, including the Correlation coefficient, defined as a product-moment; the method of moments for the fitting of distributions to samples; Pearson's system of continuous curves that forms the basis of the now conventional continuous probability distributions; Chi distance a precursor and special case of the Mahalanobis distance and P-value, defined as the probability measure of the complement of the ball with the hypothesized value as center point and chi distance as radius. He also introduced the term 'standard deviation'.

He also founded the statistical hypothesis testing theory, Pearson's chi-squared test and principal component analysis. In 1911 he founded the world's first university statistics department at University College London.

Ronald Fisher, "A genius who almost single-handedly created the foundations for modern statistical science",

The second wave of mathematical statistics was pioneered by Ronald Fisher who wrote two textbooks, *Statistical Methods for Research Workers*, published in 1925 and *The*

Design of Experiments in 1935, that were to define the academic discipline in universities around the world. He also systematized previous results, putting them on a firm mathematical footing. In his 1918 seminal paper *The Correlation between Relatives on the Supposition of Mendelian Inheritance*, the first use to use the statistical term, variance. In 1919, at Rothamsted Experimental Station he started a major study of the extensive collections of data recorded over many years. This resulted in a series of reports under the general title *Studies in Crop Variation*. In 1930 he published *The Genetical Theory of Natural Selection* where he applied statistics to evolution.

Over the next seven years, he pioneered the principles of the design of experiments and elaborated his studies of analysis of variance. He furthered his studies of the statistics of small samples. Perhaps even more important, he began his systematic approach of the analysis of real data as the springboard for the development of new statistical methods. He developed computational algorithms for analyzing data from his balanced experimental designs. In 1925, this work resulted in the publication of his first book, *Statistical Methods for Research Workers*. This book went through many editions and translations in later years, and it became the standard reference work for scientists in many disciplines. In 1935, this book was followed by *The Design of Experiments*, which was also widely used.

In addition to analysis of variance, Fisher named and promoted the method of maximum likelihood estimation. Fisher also originated the concepts of sufficiency, ancillary statistics, Fisher's linear discriminator and Fisher information. His article *On a distribution yielding the error functions of several well known statistics* (1924) presented Pearson's chi-squared test and William Gosset's t in the same framework as the Gaussian distribution, and his own parameter in the analysis of variance Fisher's z-distribution (more commonly used decades later in the form of the F distribution). The 5% level of significance appears to have been introduced by Fisher in 1925. Fisher stated that deviations exceeding twice the standard deviation are regarded as significant. Before this deviations exceeding three times the probable error were considered significant. For a symmetrical distribution the probable error is half the interquartile range. For a normal distribution the probable error is approximately 2/3 the standard deviation. It appears that Fisher's 5% criterion was rooted in previous practice.

Other important contributions at this time included Charles Spearman's rank correlation coefficient that was a useful extension of the Pearson correlation coefficient. William Sealy Gosset, the English statistician better known under his pseudonym of *Student*, introduced Student's t-distribution, a continuous probability distribution useful in situations where the sample size is small and population standard deviation is unknown.

Egon Pearson (Karl's son) and Jerzy Neyman introduced the concepts of "Type II" error, power of a test and confidence intervals. Jerzy Neyman in 1934 showed that stratified random sampling was in general a better method of estimation than purposive (quota) sampling.

Design of Experiments

James Lind carried out the first ever clinical trial in 1747, in an effort to find a treatment for scurvy.

In 1747, while serving as surgeon on HM Bark *Salisbury*, James Lind carried out a controlled experiment to develop a cure for scurvy. In this study his subjects' cases "were as similar as I could have them", that is he provided strict entry requirements to reduce extraneous variation. The men were paired, which provided blocking. From a modern perspective, the main thing that is missing is randomized allocation of subjects to treatments.

James Lind is today often described as a one-factor-at-a-time experimenter. Similar one-factor-at-a-time (OFAT) experimentation was performed at the Rothamsted Research Station in the 1840s by Sir John Lawes to determine the optimal inorganic fertilizer for use on wheat.

A theory of statistical inference was developed by Charles S. Peirce in "Illustrations of the Logic of Science" (1877–1878) and "A Theory of Probable Inference" (1883), two publications that emphasized the importance of randomization-based inference in statistics. In another study, Peirce randomly assigned volunteers to a blinded, repeated-measures design to evaluate their ability to discriminate weights.

Peirce's experiment inspired other researchers in psychology and education, which developed a research tradition of randomized experiments in laboratories and specialized textbooks in the 1800s. Peirce also contributed the first English-language publication on an optimal design for regression-models in 1876. A pioneering optimal design for polynomial regression was suggested by Gergonne in 1815. In 1918 Kirstine Smith published optimal designs for polynomials of degree six (and less).

The use of a sequence of experiments, where the design of each may depend on the results of previous experiments, including the possible decision to stop experimenting, was pioneered by Abraham Wald in the context of sequential tests of statistical hypotheses. Surveys are available of optimal sequential designs, and of adaptive designs. One specific type of sequential design is the "two-armed bandit", generalized to the multi-armed bandit, on which early work was done by Herbert Robbins in 1952.

The term "design of experiments" (DOE) derives from early statistical work performed by Sir Ronald Fisher. He was described by Anders Hald as "a genius who almost single-handedly created the foundations for modern statistical science." Fisher initiated the principles of design of experiments and elaborated on his studies of "analysis of variance". Perhaps even more important, Fisher began his systematic approach to the analysis of real data as the springboard for the development of new statistical methods. He began to pay particular attention to the labour involved in the necessary computations performed by hand, and developed methods that were as practical as they were founded in rigour. In 1925, this work culminated in the publication of his first book, *Statistical Methods for Research Workers*. This went into many editions and translations in later years, and became a standard reference work for scientists in many disciplines.

A methodology for designing experiments was proposed by Ronald A. Fisher, in his innovative book *The Design of Experiments* (1935) which also became a standard. As an example, he described how to test the hypothesis that a certain lady could distinguish by flavour alone whether the milk or the tea was first placed in the cup. While this sounds like a frivolous application, it allowed him to illustrate the most important ideas of experimental design.

Agricultural science advances served to meet the combination of larger city populations and fewer farms. But for crop scientists to take due account of widely differing geographical growing climates and needs, it was important to differentiate local growing conditions. To extrapolate experiments on local crops to a national scale, they had to extend crop sample testing economically to overall populations. As statistical methods advanced (primarily the efficacy of designed experiments instead of one-factor-at-a-time experimentation), representative factorial design of experiments began to enable the meaningful extension, by inference, of experimental sampling results to the population as a whole. But it was hard to decide how representative was the crop sample chosen. Factorial design methodology showed how to estimate and correct for any random variation within the sample and also in the data collection procedures.

Bayesian Statistics

Pierre-Simon, marquis de Laplace, one of the main early developers of Bayesian statistics.

The term *Bayesian* refers to Thomas Bayes (1702–1761), who proved a special case of what is now called Bayes' theorem. However it was Pierre-Simon Laplace (1749–1827) who introduced a general version of the theorem and applied it to celestial mechanics, medical statistics, reliability, and jurisprudence. When insufficient knowledge was available to specify an informed prior, Laplace used uniform priors, according to his "principle of insufficient reason". Laplace assumed uniform priors for mathematical simplicity rather than for philosophical reasons. Laplace also introduced primitive versions of conjugate priors and the theorem of von Mises and Bernstein, according to which the posteriors corresponding to initially differing priors ultimately agree, as the number of observations increases. This early Bayesian inference, which used uniform priors following Laplace's principle of insufficient reason, was called "inverse probability" (because it infers backwards from observations to parameters, or from effects to causes).

After the 1920s, inverse probability was largely supplanted by a collection of methods that were developed by Ronald A. Fisher, Jerzy Neyman and Egon Pearson. Their methods came to be called frequentist statistics. Fisher rejected the Bayesian view, writing that "the theory of inverse probability is founded upon an error, and must be wholly rejected". At the end of his life, however, Fisher expressed greater respect for the essay of Bayes, which Fisher believed to have anticipated his own, fiducial approach to probability; Fisher still maintained that Laplace's views on probability were "fallacious rubbish". Neyman started out as a "quasi-Bayesian", but subsequently developed confidence intervals (a key method in frequentist statistics) because "the whole theory would look nicer if it were built from the start without reference to Bayesianism and priors". The word *Bayesian* appeared around 1950, and by the 1960s it became the term preferred by those dissatisfied with the limitations of frequentist statistics.

In the 20th century, the ideas of Laplace were further developed in two different directions, giving rise to *objective* and *subjective* currents in Bayesian practice. In the objectivist stream, the statistical analysis depends on only the model assumed and the data analysed. No subjective decisions need to be involved. In contrast, "subjectivist" statisticians deny the possibility of fully objective analysis for the general case.

In the further development of Laplace's ideas, subjective ideas predate objectivist positions. The idea that 'probability' should be interpreted as 'subjective degree of belief in a proposition' was proposed, for example, by John Maynard Keynes in the early 1920s. This idea was taken further by Bruno de Finetti in Italy (*Fondamenti Logici del Ragionamento Probabilistico*, 1930) and Frank Ramsey in Cambridge (*The Foundations of Mathematics*, 1931). The approach was devised to solve problems with the frequentist definition of probability but also with the earlier, objectivist approach of Laplace. The subjective Bayesian methods were further developed and popularized in the 1950s by L.J. Savage.

Objective Bayesian inference was further developed by Harold Jeffreys at the University of Cambridge. His seminal book "Theory of probability" first appeared in 1939 and played an important role in the revival of the Bayesian view of probability. In 1957, Edwin Jaynes promoted the concept of maximum entropy for constructing priors, which is an important principle in the formulation of objective methods, mainly for discrete problems. In 1965, Dennis Lindley's 2-volume work "Introduction to Probability and Statistics from a Bayesian Viewpoint" brought Bayesian methods to a wide audience. In 1979, José-Miguel Bernardo introduced reference analysis, which offers a general applicable framework for objective analysis. Other well-known proponents of Bayesian probability theory include I.J. Good, B.O. Koopman, Howard Raiffa, Robert Schlaifer and Alan Turing.

In the 1980s, there was a dramatic growth in research and applications of Bayesian methods, mostly attributed to the discovery of Markov chain Monte Carlo methods, which removed many of the computational problems, and an increasing interest in nonstandard, complex applications. Despite growth of Bayesian research, most undergraduate teaching is still based on frequentist statistics. Nonetheless, Bayesian methods are widely accepted and used, such as for example in the field of machine learning.

Important Contributors to Statistics

• Thomas Bayes	• Carl Friedrich Gauss	• Blaise Pascal
• George E. P. Box	• William Sealey Gosset ("Student")	• Karl Pearson
• Pafnuty Chebyshev	• Andrey Kolmogorov	• Charles S. Peirce
• David R. Cox	• Pierre-Simon Laplace	• Adolphe Quetelet
• Gertrude Cox	• Erich L. Lehmann	• C. R. Rao
• Harald Cramér	• Aleksandr Lyapunov	• Walter A. Shewhart
• Francis Ysidro Edgeworth	• Anil Kumar Gain	• Charles Spearman
• Bradley Efron	• Prasanta Chandra Mahalanobis	• Charles Stein
• Bruno de Finetti	• Abraham De Moivre	• Thorvald N. Thiele
• Ronald A. Fisher	• Jerzy Neyman	• John Tukey
• Francis Galton	• Florence Nightingale	• Abraham Wald

Founders of Statistics

Statistics is the theory and application of mathematics to the scientific method including hypothesis generation, experimental design, sampling, data collection, data summarization, estimation, prediction and inference from those results to the population from which the experimental sample was drawn. This article lists statisticians who have been instrumental in the development of theoretical and applied statistics.

Name	Nation-ality	Birth	Death	Contribution
Al-Kindi	Iraqi	801	873	Developed the first code breaking algorithm based on frequency analysis. He wrote a book entitled "Manuscript on Deciphering Cryptographic Messages", containing detailed discussions on statistics
Graunt, John	English	1620	1674	Pioneer of demography who produced the first life table
Bayes, Thomas	English	1702	1761	Developed the interpretation of probability now known as Bayes theorem
Laplace, Pierre-Simon	French	1749	1827	Co-invented Bayesian statistics. Invented exponential families (Laplace transform), conjugate prior distributions, asymptotic analysis of estimators (including negligibility of regular priors). Used maximum-likelihood and posterior-mode estimation and considered (robust) loss functions
Playfair, William	Scottish	1759	1823	Pioneer of statistical graphics
Gauss, Carl Friedrich	German	1777	1855	Invented least squares estimation methods (with Legendre). Used loss functions and maximum-likelihood estimation
Quetelet, Adolphe	Belgian	1796	1874	Pioneered the use of probability and statistics in the social sciences
Nightingale, Florence	English	1820	1910	Applied statistical analysis to health problems, contributing to the establishment of epidemiology and public health practice. Developed statistical graphics especially for mobilizing public opinion. First female member of the Royal Statistical Society.
Galton, Francis	English	1822	1911	Invented the concepts of standard deviation, correlation, regression
Thiele, Thorvald N.	Danish	1838	1910	Introduced cumulants and the term "likelihood". Introduced a Kalman filter in time-series
Peirce, Charles Sanders	American	1839	1914	Formulated modern statistics in "Illustrations of the Logic of Science" (1877–1878) and "A Theory of Probable Inference" (1883). With a repeated measures design, introduced blinded, controlled randomized experiments (before Fisher). Invented optimal design for experiments on gravity, in which he "corrected the means". He used correlation, smoothing, and improved the treatment of outliers. Introduced terms "confidence" and "likelihood" (before Neyman and Fisher). While largely a frequentist, Peirce's possible world semantics introduced the "propensity" theory of probability.
Edgeworth, Francis Ysidro	Irish	1845	1926	Revived exponential families (Laplace transforms) in statistics. Extended Laplace's (asymptotic) theory of maximum-likelihood estimation. Introduced basic results on information, which were extended and popularized by R. A. Fisher

Name	Nationality	Birth	Death	Contribution
Pearson, Karl	English	1857	1936	Numerous innovations, including the development of the Pearson chi-squared test and the Pearson correlation. Founded the Biometrical Society and Biometrika, the first journal of mathematical statistics and biometry
Spearman, Charles	English	1863	1945	Extended the Pearson correlation coefficient to the Spearman's rank correlation coefficient
Gosset, William Sealy (known as "Student")	English	1876	1937	Discovered the Student t distribution and invented the Student's t-test
Fisher, Ronald	English	1890	1962	Wrote the textbooks and articles that defined the academic discipline of statistics, inspiring the creation of statistics departments at universities throughout the world. Systematized previous results with informative terminology, substantially improving previous results with mathematical analysis (and claims). Developed the analysis of variance, clarified the method of maximum likelihood (without the uniform priors appearing in some previous versions), invented the concept of sufficient statistics, developed Edgeworth's use of exponential families and information, introducing *observed* Fisher information, and many theoretical concepts and practical methods, particularly for the design of experiments
Bonferroni, Carlo Emilio	Italian	1892	1960	Invented the Bonferroni correction for multiple comparisons
Wilcoxon, Frank	Irish-American	1892	1965	Invented two statistical tests: Wilcoxon rank-sum test and the Wilcoxon signed-rank test
Neyman, Jerzy	Polish-American	1894	1981	Discovered the confidence interval and co-developed the Neyman–Pearson lemma
Deming, W. Edwards	American	1900	1993	Developed methods for statistical quality control
Pearson, Egon	English	1895	1980	Co-developed the Neyman–Pearson lemma of statistical hypothesis testing
Finetti, Bruno de	Italian	1906	1985	Pioneer of the "operational subjective" conception of probability. Used this as the basis for exposition of the Bayesian method of statistical analysis. Developed the representation theorem for exchangeable random variables showing that they are the basis of the IID model in statistics.
Kendall, Maurice	English	1907	1983	Co-developed methods for assessing statistical randomness; invented Kendall tau rank correlation coefficient
Tukey, John	American	1915	2000	Jointly popularized Fast Fourier transformation, pioneer of exploratory data analysis and graphical presentation of data, developed the jackknife for variance estimation, invented the box plot.
Blackwell, David	American	1919	2010	Co-developed Rao-Blackwell theorem and wrote one of the first Bayesian textbooks, *Basic Statistics*.

Name	Nationality	Birth	Death	Contribution
Rao, Calyampudi Radhakrishna	Indian	1920		Co-developed Cramér–Rao bound and Rao–Blackwell theorem, invented MINQUE method of variance component estimation.
Cox, David	English	1924		Developed the proportional hazards model for the analysis of survival data
Efron, Bradley	American	1938		Invented the bootstrap resampling technique for deriving an empirical distribution of an estimate of a model parameter

Founders of Departments of Statistics

The role of a department of statistics is discussed in a 1949 article by Harold Hotelling, which helped to spur the creation of many departments of statistics.

Year	Country	University	Founder
1911	England	University College London	Pearson, Karl
1918	United States	Department of Biostatistics, Johns Hopkins Bloomberg School of Public Health	Pearl, Raymond
~1931	India	Indian Statistical Institute	Prasanta Chandra Mahalanobis
~1931	United States	Columbia University	Hotelling, Harold
1933	USA	Iowa State University	Snedecor, George W.
1941	USA	North Carolina State University	Cox, Gertrude
1942	Sweden	Uppsala University	Wold, Herman
1947	England	University of Manchester	Bartlett, M. S.
1947	USA	Department of Biometry and Statistics, Cornell University	Federer, Walter T.
1948	USA	Stanford University	—
1948	India	University of Mumbai	M. C. Chakrabarti
1949	USA	University of North Carolina at Chapel Hill	—
1949	USA	University of Chicago	—
1953	England	Cambridge University, Statistics Lab	Wishart, John
1953	India	University of Pune	V. S. Huzurbazar
1955	USA	University of California, Berkeley	Neyman, Jerzy
1957	USA	Harvard University	Cochran, W. G. Mosteller, Frederick
1957	Australia	University of Sydney	Lancaster, H.O.

Year	Country	University	Founder
1962	USA	Texas A&M University	Hartley, Herman Otto
1963	USA	Yale University	Anscombe, Francis
1965	USA	Princeton University	Tukey, John W
1965	USA	University of Iowa	Hogg, Robert V.
1966	Scotland	University of Glasgow	Aitchison, John Silvey, David
1973	USA	The Ohio State University	Whitney, D. Ransom
1979	Canada	University of Toronto	—
1981	India	Vidyasagar University	Anil Kumar Gain
1982	Hong Kong	Chinese University of Hong Kong	Tong, Howell
1984	India	Banaras Hindu University	Singh, S.N.
1988	England	University of Oxford	Hinkley, D. V.
1996	USA	University of Virginia School of Medicine	Harrell, Frank E.

References

- Zacks, S. (1996) "Adaptive Designs for Parametric Models". In: Ghosh, S. and Rao, C. R., (Eds) (1996). "Design and Analysis of Experiments," Handbook of Statistics, Volume 13. North-Holland. ISBN 0-444-82061-2. (pages 151–180)

- Edwards, A.W.F. (2005). "R. A. Fisher, Statistical Methods for Research Workers, 1925". In Grattan-Guinness, Ivor. Landmark writings in Western mathematics 1640-1940. Amsterdam Boston: Elsevier. ISBN 9780444508713.

- Bernardo J (2005). "Reference analysis". Handbook of statistics. Handbook of Statistics. 25: 17–90. doi:10.1016/S0169-7161(05)25002-2. ISBN 9780444515391.

- Singh, Simon (2000). The code book : the science of secrecy from ancient Egypt to quantum cryptography (1st Anchor Books ed.). New York: Anchor Books. ISBN 0-385-49532-3.

- Wermuth, Nanny; Anthony C. Davison; Dodge, Yadolah (2005). Celebrating Statistics: Papers in honour of Sir David Cox on his 80th birthday (Oxford Statistical Science Series). Oxford University Press. ISBN 0-19-856654-9. OCLC 185035518.

- Dixit, Ulhas J.; Satam, Meena R. (1999). Dedicated to MC Chakrabarti, Book (Papers) - Statistical Inference and Design of Experiments. Narosa Publishing House. ISBN 978-81-7319-281-4.

- Klaus Hinkelmann (2012). Design and Analysis of Experiments, Special Designs and Applications. John Wiley & Sons. p. xvii.

- David Leonhardt (28 July 2000). "John Tukey, 85, Statistician; Coined the Word 'Software'". New York Times. Retrieved 2 May 2010.

Permissions

All chapters in this book are published with permission under the Creative Commons Attribution Share Alike License or equivalent. Every chapter published in this book has been scrutinized by our experts. Their significance has been extensively debated. The topics covered herein carry significant information for a comprehensive understanding. They may even be implemented as practical applications or may be referred to as a beginning point for further studies.

We would like to thank the editorial team for lending their expertise to make the book truly unique. They have played a crucial role in the development of this book. Without their invaluable contributions this book wouldn't have been possible. They have made vital efforts to compile up to date information on the varied aspects of this subject to make this book a valuable addition to the collection of many professionals and students.

This book was conceptualized with the vision of imparting up-to-date and integrated information in this field. To ensure the same, a matchless editorial board was set up. Every individual on the board went through rigorous rounds of assessment to prove their worth. After which they invested a large part of their time researching and compiling the most relevant data for our readers.

The editorial board has been involved in producing this book since its inception. They have spent rigorous hours researching and exploring the diverse topics which have resulted in the successful publishing of this book. They have passed on their knowledge of decades through this book. To expedite this challenging task, the publisher supported the team at every step. A small team of assistant editors was also appointed to further simplify the editing procedure and attain best results for the readers.

Apart from the editorial board, the designing team has also invested a significant amount of their time in understanding the subject and creating the most relevant covers. They scrutinized every image to scout for the most suitable representation of the subject and create an appropriate cover for the book.

The publishing team has been an ardent support to the editorial, designing and production team. Their endless efforts to recruit the best for this project, has resulted in the accomplishment of this book. They are a veteran in the field of academics and their pool of knowledge is as vast as their experience in printing. Their expertise and guidance has proved useful at every step. Their uncompromising quality standards have made this book an exceptional effort. Their encouragement from time to time has been an inspiration for everyone.

The publisher and the editorial board hope that this book will prove to be a valuable piece of knowledge for students, practitioners and scholars across the globe.

Index

www.ingramcontent.com/pod-product-compliance
Lightning Source LLC
Chambersburg PA
CBHW061935190326
41458CB00009B/2741